Oktober 86

FOREST SITE AND PRODUCTIVITY

FORESTRY SCIENCES

Baas P, ed: New Perspectives in Wood Anatomy. 1982. ISBN 90-247-2526-7

Prins CFL, ed: Production, Marketing and Use of Finger-Jointed Sawnwood. 1982. ISBN 90-247-2569-0

Oldeman RAA, et al., eds: Tropical Hardwood Utilization: Practice and Prospects. 1982. ISBN 90-247-2581-X

Den Ouden P and Boom BK: Manual of Cultivated Conifers: Hardy in Cold and Warm-Temperate Zone. 1982. ISBN 90-247-2148-2 paperback; ISBN 90-247-2644-1 hardbound.

Bonga JM and Durzan DJ, eds: Tissue Culture in Forestry. 1982. ISBN 90-247-2660-3

Satoo T and Magwick HAI: Forest Biomass. 1982. ISBN 90-247-2710-3

Van Nao T, ed: Forest Fire Prevention and Control. 1982. ISBN 90-247-3050-3

Douglas J: A Re-appraisal of Forestry Development in Developing Countries. 1983. ISBN 90-247-2830-4

Gordon JC and Wheeler CT, eds: Biological Nitrogen Fixation in Forest Ecosystems: Foundations and Applications. 1983. ISBN 90-247-2849-5

Hummel FC, ed: Forest Policy: A Contribution to Resource Development. 1984. ISBN 90-247-2883-5

Duryea ML and Landis TD, eds: Forest Nursery Manual: Production of Bareroot Seedlings. 1984. ISBN 90-247-2913-0

Manion PD, ed: Scleroderris Canker of Conifers. 1984. ISBN 90-247-2912-2

Staaf KAG and Wiksten NA: Tree Harvesting Techniques. 1984. ISBN 90-247-2994-7

Duryea ML and Brown GN, eds: Seedling Physiology and Reforestation Success. 1984. ISBN 90-247-2949-1

Boyd JD: Biophysical Control of Microfibril Orientation in Plant Cell Walls. 1985. ISBN 90-247-3101-1

Findlay WPK, ed: Preservation of Timber in the Tropics. 1985. ISBN 90-247-3112-7

Samset I: Winch and Cable Systems. 1985. ISBN 90-247-3205-0

Leary RA: Interaction Theory in Forest Ecology and Management. 1985. ISBN 90-247-3220-4

Gessel SP: Forest Site and Productivity. 1986. ISBN 90-247-3284-0

Forest site and productivity

edited by

S.P. GESSEL
College of Forest Resources
University of Washington
Seattle, Washington, USA

1986 **MARTINUS NIJHOFF PUBLISHERS**
a member of the KLUWER ACADEMIC PUBLISHERS GROUP
DORDRECHT / BOSTON / LANCASTER

Distributors

for the United States and Canada: Kluwer Academic Publishers, 190 Old Derby Street, Hingham, MA 02043, USA
for the UK and Ireland: Kluwer Academic Publishers, MTP Press Limited, Falcon House, Queen Square, Lancaster LA1 1RN, UK
for all other countries: Kluwer Academic Publishers Group, Distribution Center, P.O. Box 322, 3300 AH Dordrecht, The Netherlands

Library of Congress Cataloging in Publication Data

ISBN 90-247-3284-0 (this volume)

PRINTED IN THE NETHERLANDS

CONTENTS

PREFACE

A knowledge of forest site and forest productivity variables is fundamental to sound forest practice everywhere. The ability to identify sites and site problems correctly and manipulate productivity variables for maintenance or improvement of productivity is the basis of modern forest management. Although the basic facts regarding forest site and productivity apply throughout the world, the application of information and the response to manipulation vary greatly and depend on local forest conditions.

The September 1981 World Congress of the International Union of Forest Research Organizations (IUFRO) in Kyoto, Japan was the occasion for the special meeting on Forest Site and Productivity sponsored by the IUFRO Site Group S1.02. This meeting brought together forest site and productivity researchers from across the world to review current thought and the state of site research.

Information not ordinarily available in one place was presented at this meeting. As organizer of the session, I decided to attempt to publish the papers in one volume. Arrangements were made with a publisher, Martinus Nijhoff, and also with the authors. The process of publication has taken longer than desirable, but the volume does appear at an opportune time coincident with the 1986 IUFRO World Congress in Yugoslavia. Material contained in this publication will set the stage for Site Group discussions at the 1986 meeting.

This volume assembles the thought of forest research workers from many different countries and therefore many different kinds of forests. The papers are organized under the headings "Site and Land Classification," "Nutrient Cycling and Site Productivity," and "Fertilization and Site Growth Response."

I believe the reader will find the writings useful in terms of both research planning and application to forestry.

Stanley P. Gessel
University of Washington

Part 1

SITE AND LAND CLASSIFICATION

1

THE RELATIONSHIP BETWEEN SITE CLASSIFICATION AND TERRAIN CLASSIFICATION

H. Löffler

THE PROBLEM

With a certain abstraction one can define site classification (or site mapping) as a means of grouping forest sites according to their capability of growing trees and terrain classification as a means of grouping forest land according to ease or difficulty of forest operations. Whether the so-called hazard rating or susceptibility classification, concerning mainly the risk of soil compaction and mass movement, should be dealt with as an issue of site or terrain classification or as a separate subject is not a question of this contribution.

Site classification is an important and extremely valuable aid to forest management. As to terrain classification, experience shows that the information it provides becomes increasingly necessary (1) with rising mechanization (rising input of capital and energy), (2) with the extension of forest operations to hitherto inaccessible areas or areas hardly explored up to now, and (3) under the growing urgency of demands that operational methods and equipment should guarantee high economic efficiency as well as environmental harmlessness.

Compared with site classification, terrain classification is a fairly recent development. As a scientific, systematic approach it was not started before the early 1950s. The Nordic countries and Great Britain have the greatest experience, but in several other countries terrain classification systems are in preparation (see for example, Skogsarbeten 1969; Samset 1975; Rowan 1977; Nilsson 1979; Sutton 1979).

For about two decades international institutions--in particular IUFRO, Joint FAO/ECE/ILO Committee, and FAO--have launched attempts at working out at least comparability of the different terrain classification systems (see, for example, von Segebaden et al. 1967; FAO 1973; Rowan 1974). At present a working group consisting of members of the Joint Committee and IUFRO is making a new move in this direction.

Hitherto the development and activities of site classification and terrain classification have run along parallel, separate lines, even though both approaches have the same objective--merely a different perspective. Since data collecting for site and terrain classification requires considerable expense, it seems reasonable to ask whether there may be a chance for a common procedure.

Principally the following prospects are imaginable: (1) Site classification provides or is altered to provide the information rendered by terrain classification. Some experts of site classification are indeed of the opinion that terrain classification is superfluous and site classification could meet all corresponding demands. (2) Terrain classification and site classification continue to be treated as separate approaches, but efforts are made to exchange basic data that are required for both classifications. The optimal solution would be joint data collecting and consequently a common data base that would allow the derivation of both site and terrain classification.

CHARACTERIZATION OF TERRAIN FROM AN OPERATIONAL STANDPOINT

One has to distinguish between (1) a descriptive or primary terrain classification that describes and classifies forest land independently from the possibilities and limitations of machines and operational methods and (2) functional or secondary terrain classifications characterizing an area with regard to distinct equipment and methods. In this context the discussion is confined to the descriptive terrain classification.

It has been recommended to distinguish two classification levels: a macrodescriptive or reconnaissance survey level and a microdescriptive or detailed survey level. Classification on the macrodescriptive level will serve to characterize larger land units with regard to macrotopography (landform, geomorphological patterns), infrastructure (transportation system), and--when applied exclusively (i.e., not combined with the detailed level)--patterns of soil. Classification on the microdescriptive level is generally provided for small land units, sample plots, working sites, and so forth. Table 1 gives a general survey of the terrain features required for the two levels.

In the following we shall deal with the microdescriptive level. On that level, terrain conditions will be expressed by terrain classes, which are assembled by predetermined categories of single parameters (ground conditions, slope conditions, ground roughness, and possibly average off-road skidding distance).

Slope, ground roughness (or microtopography), and off-road transportation distance are usually not considered in site classification, or at least not to the degree required in terrain classification. Therefore, site classification in its present structure cannot replace terrain classification unless it is widened by additional information. Besides, the mapping units of site classification and terrain classification are generally incongruent, owing to different delineation criteria and different importance attributed to the criteria used in the two approaches.

GROUND CONDITIONS

It is the factor "ground [soil] conditions" that mainly forms the point of contact between the two classification procedures. What is meant by "ground conditions" in terrain classification? In Table 2 the parameters used to characterize ground conditions in respect to forest operations are listed. Their relative importance is indicated by number of crosses, with soil strength claiming by far the greatest interest. Soil strength may be defined as "the ability of a soil body to bear or withstand stresses without collapsing or deforming excessively" (Hillel 1980). One can state that terrain classification is aimed at grouping forest sites according to their supporting capacity as far as ground conditions are concerned.

Table 1. Classification levels and terrain features.

Terrain Features	Classification Level: Macrodescription obligatory	Macrodescription optional	Microdescription obligatory	Microdescription optional
Macrotopography (landform, patterns of slope), including range of altitude	X			
Drainage system		X		
Climate (annual precipitation, average snow conditions, mean annual temperature, etc.)	X			
Geology, parent material, possibly geogenesis	X			
Forest infrastructure	X			
Public infrastructure (transportation system)		X		
Predominant (average) ground conditions (patterns of soil)	X (when macrodescription is used exclusively)			
Ground conditions			X	
Slope conditions			X	
Ground roughness			X	
Average off-road skidding distance				X

Information on soil strength can be gained in two ways: (1) by direct measuring, and (2) by assessment based on statistical relationship.

Apart from methodological problems, the direct recording of soil strength is an expensive and complicated task. Therefore, direct soil strength measurements are rare in forestry. Considering, however, the importance of a better knowledge about the soil supporting capacity, direct measurement ought to be intensified, although it is unlikely that a sufficient amount of data will soon be available in order to get ahead with terrain classification.

Several investigations point to a stochastic relationship between soil strength, regardless of the kind of measurement, and the properties

Table 2. Parameters to characterize ground conditions within a descriptive terrain classification system.

	Parameter	Importance
1.	<u>Soil Texture</u>	
	- mineral soils (≥ 70% mineral components): grain size distribution (percentage gravel, sand, silt, and clay; sieve curve) - organic soils (≥ 30% organic components): terms of wetland classification (muskeg, peat, etc.)	+ + +
	- vertical homogeneity	+
2.	<u>Moisture Conditions</u> (prevailing or "normal")	
	- soil moisture content, or - site moisture regime	+ + +
	- soil drainage conditions	+
3.	<u>Soil Dry Density</u>	+ +
4.	<u>Plastic Properties</u> (cohesive soils)	
	- liquid limit, plastic limit	+ + +
	- plasticity index	+ + +
	- activity of clay	+ +
5.	<u>Soil Depth</u> (depth of unconsolidated material)	
	- mineral soils: from surface down to bedrock or to impermeable pan - organic soils: from surface down to mineral layer	+ +
6.	<u>Stoniness</u> (content of particles > 60 mm)	+ +
7.	<u>Strengthening Factors</u>	
	- stumps (roots)	+
	- slash cover	+
8.	<u>Soil Strength</u> (supporting capacity at prevailing or "normal" moisture conditions)	+ + + +
	- analytical, parametric method (Bevameter values) - empirical methods: California Bearing Ratio, modulus of elasticity, Cone Index, etc.	

+ + + + very high
+ + + high
+ + medium
+ low

recorded under numbers 1 to 7 in Table 2. If further investigations confirm and specify these correlations, the description and classification of ground conditions within terrain classification could be based on the properties mentioned--with soil texture, moisture conditions, soil dry density, and plastic properties considered as the most important ones.

Modern site classifications are as a rule hierarchically structured. On the higher (macrodescriptive) classification levels, climate and landform patterns are mainly used as grouping criteria. On the lower levels, particularly on the lowest level, soil properties play a leading part in the delineation of mapping units. Although the ground conditions in a site classification are characterized by other features than in a terrain classification, it may be expected that the technically oriented ground conditions are more or less homogenous also in a site unit.

Thus, in order to be able to utilize site classification for purposes of terrain classification, an interpretation of the site classification units would be necessary at least with regard to soil texture, soil moisture condition, soil dry density, and--for cohesive soils--plastic properties.

According to preliminary results of investigation in the Federal Republic of Germany, the "site type" (the smallest mapping unit of the German site classification system) guarantees an excellent discriminating effect as to the parameters mentioned.

OUTLOOK AND RECOMMENDATIONS

It must be regretted that site classification and terrain classification have been treated almost in isolation from each other and that mutual understanding between these two disciplines is quite often only minute. In cases where site classification is already completed and terrain classification is under preparation, one should endeavor to utilize information on soil conditions as indicated above. Where both classification procedures still have to be executed, a joint data collecting should be attempted. Above all, the experts of site and terrain classification should work together.

REFERENCES

FAO. 1973. Tentative checklist for describing and quantifying logging, transport and road conditions.

Hillel, D. 1980. Fundamentals of soil physics. Academic Press, New York.

Nilsson, G. 1979. The terrain classification system for stand establishment operations in the Nordic countries. Pre-edition Symposium on Stand Establishment, VNIILM PUSHKINO.

Rowan, A. A. 1974. A review of terrain and operational systems. FAO.

_________. 1977. Terrain classification. Forestry Commission, Forest Record 114.

Samset, I. 1975. The accessibility of forest terrain and its influence on forestry conditions in Norway. NISK Report 32.1.

von Segebaden, G., R. Strømnes, and H. I. Winer. 1967. Proposal for international system of terrain classification. XIV IUFRO World Congress, Munich, Division 3.

Skogsarbeten. 1969. Terrain classification for Swedish forestry. Forskningsstiftelsen Skogsarbeten, Report No. 9.

Sutton, A. R. 1979. Terrain classification in Great Britain. Pre-edition Symposium on Stand Establishment, VNIILM PUSHKINO.

2

THE GEOMORPHIC APPROACH TO SITE DELINEATION IN INTENSIVELY MANAGED EXOTIC CONIFER PLANTATIONS

D. C. Grey

INTRODUCTION

A rational system for the subdivision of land is a prerequisite for efficient planning and management of any forestry enterprise. The land surface reflects a balance between past processes and the physical, chemical, and hydrological properties of the underlying materials. Geometry of the land surface has considerable influence on currently operating processes; for example, slope is an important property which determines soil and regolith stability, and the interaction of slope and aspect accounts for a large proportion of the shortwave radiation received in any local area. Surface configuration plays a dominant role in the moisture status and temperature regime of the solum.

Interest in land-surface geometry diminishes at the very small and very large scales, because at these extremes the natural characteristics of the surface become a relatively unimportant factor in general land use. Planning at national or regional levels is chiefly concerned with social, economic, and strategic considerations. In this context the terrain plays only a small part. Likewise, at large scales, the construction of major roads and buildings or the planning of intensive urban and industrial development involves such great effort and economic input that land-surface considerations, including possible modifications, form an insignificant proportion of the total cost. The domain of land-surface characterization is therefore limited to the intermediate levels of land use, such as managed exotic conifer plantations.

Land features that can be identified and mapped at scales of 1:2,000 to 1:60,000 serve as a useful guide for site recognition, which involves the collection, analysis, and presentation of all relevant data concerning the landscape geometry and surficial materials--in particular, their location and spatial relations. Site data may be obtained by field observation or by measurement and synthesis from aerial photographs and topographic maps.

THE LAND SURFACE

Each point on the earth's surface has a number of interrelated attributes that, if considered in sufficient detail, make that point unique. Qualities associated with the land surface differ in their spatial expression. Some features have a clearly expressed form, such as a cliff face or stream bank; others, such as gradual slope changes, are diffuse and difficult to map (Mitchell 1973).

There is an intrinsic advantage in assessing the land surface in terms of interacting units. Any site receives water, chemicals, soil, and seeds from higher areas in the landscape and sheds these in turn to other sites lower down. Trees growing on ridge crests are subject to greater exposure to wind, orographic rainfall, and evapotranspiration than trees growing in the mesic, valley cove site.

Perception of landforms is influenced by observer position, distance (scale), form, space, sequence, and time factors (Litton 1972). Observer position defines the field of view; there is a likelihood of screening, enclosure, or optical distortion, which applies equally to field observation or to stereoscopic images. On uniform slopes, a plunging view appears steeper than the ascending view. The foreground enhances clarity, simplicity, and detail, but there is a lack of perspective. In viewing from a distance, emphasis is placed on edge effects, outline, and tonal contrasts. Convex shapes emphasize form while concave segments convey the concept of space (they appear larger). It is necessary to train the eye movements and to order the sequence with which the landscape is viewed, in order to avoid bias in judgment. Shadows, cloud patterns, seasonal color changes, and even wind motion can influence perception of the landscape.

Surface features may be defined on the basis of their geometric form, but be recognized by such incidental properties as texture, tone, absence of drainage patterns, or association with certain vegetation types. This has been termed recognizability. The extent to which surface properties and processes are correlated with certain landforms is reproducibility (Beckett 1974). Some degree of reproducibility is required for the transfer of knowledge from known or measured sites to new sites. To prevent a proliferation of sites and to aid classification, it is essential that equal weight be given to recognizability (the ability to map) and reproducibility (internal homogeneity and association with tree growth or other classification objectives). There is an inverse relation between recognizability and reproducibility. These inherent complexities of the landscape need to be understood and simplified if the characteristics of different sites are to be compared.

THE MODEL

Landscapes can be classified by the subdivision of extensive areas or by aggregation from smaller units. Linton (1951) recognized "flats and slopes" as landscape units "indivisible on the basis of form." Borne (1931) is usually regarded as the pioneer of the geomorphic approach to site delineation, and his suggestion that sites are "areas which for all practical purposes had similar physiography, geology and edaphic factors" is viewed as the forerunner of hierarchial land classification systems (Howard and Mitchell 1980).

Morphological mapping techniques were developed by Savigear (1965), who recognized five basic geometric forms. These forms could be separated one from the other in the landscape by slope breaks, slope changes, or inflections. Subsequently, hillslope models were proposed to facilitate the organization and condensation of Linton's and Savigear's ideas into a finite number of slope categories (Conacher and Dalrymple 1977; Young 1975; Ruhe and Walker 1968; King 1967).

These models were reorganized by the author into a useful framework for the delineation of forest sites (Grey 1979, 1980). The geomorphic units given below do not represent primary entities of the landscape. They are defined on the basis of characteristic slope and form, rather than on process and response, as are the units of Conacher and Dalrymple

(1977). All the units do not necessarily occur within low relief; some units may be extensive, and slope changes as small as half a degree could signal important site differences (Wright 1973).

Geomorphic units are given numbers in preference to names to avoid the connotations associated with previous studies of slope profile development and evolution.

Unit 1

Convex in profile, this unit is found on the divides and interfluves between local thalwegs. Slopes vary from 0° to 3°. Convexity in profile is slight: 1°-10°/100 m. The size and plan form of the unit vary greatly, for they are dependent on local relief and drainage network development. The unit forms part of the waxing or crest slope of King (1967) and the summit of Ruhe and Walker (1968), and is equivalent to land surface unit one of Conacher and Dalrymple (1977).

The soil surface shows little sign of erosion, except in semiarid areas where the removal of fines due to wind action may be important. The soils are mainly lithosolic, but may obtain considerable profile development in a humid or subhumid climate if the underlying rock is resistant to weathering. Because of its position in the landscape, this unit is exposed to wind and has a high rate of evapotranspiration. It is seldom found on extensive plateaus or within large alluvial plains and attains maximum development in areas of high local relief and denudational topographies. Unit 1 is separated from the convex unit 2 by an irregular but consistent increase in downslope convexity.

Trees on unit 1 sites show greater taper, are subject to snow damage and windfalls, and tend to have low site indices. Total volume is not always affected, unless the trees are subject to drought stress due to the seasonality of rainfall.

Unit 2

This geomorphic unit is characterized by its convexity of profile, commonly 10°-100°/100 m. It may occur as the divide in some landscapes, but is then usually dissected by incipient stream headwaters, shallow rills, or depressions, which give rise to the convexity. The macroslope varies from 3° to 5°. Terracetts and small slip scars, 2 to 6 m across, are common on argillaceous rocks.

Soil creep, the formation of leached horizons due to lateral water movement (E horizon), soft plinthite, stonelines, and discrete G horizons are common features. The soils may be deeper than those of unit 1, but the presence of perched water, drift lines, and planosolic soil processes reduces the strength properties and prevents deep rooting.

Unit 2 sites are generally narrow, sinuous, and discrete in the landscape. Timber yields are low because of the excessive windfalls, which reduce the stocking, and alternative periods of moisture surplus and deficit that hinder root development. Road banks within this unit have a tendency to fail because of undercutting. This unit corresponds to the shoulder of Ruhe and Walker, may form part of the crest and pediment elements of King, and is a combination of landsurface units two and three of Conacher and Dalrymple.

Unit 3

Steep slopes (14°-40°), irregular surface, and lack of continuous soil cover set this unit apart from all others in the landscape. Large-scale soil movement is found on this unit; landslides, mudflows, and slumping are typical features, showing that unit 3 is inherently unstable. It is associated with erosional escarpments and lithologies that differ sharply in their resistance to weathering.

The macroprofile form of unit 3 is linear but highly irregular, and all possible slope shapes have been observed in minor areas. Unit 3 is associated with the segment of the total slope profile that has maximum slope (Young 1975). Soils vary greatly in depth, but are poorly leached, because of their general youth (Inceptisols).

The backslope of Ruhe and Walker, free face element of King, and landsurface units four and five of Conacher and Dalrymple have features in common with unit 3. Tree growth is extremely erratic. In most plantations, unit 3 is seldom planted because of exploitation difficulties and the need to prevent soil erosion and landslides. An appreciable number of boulders may occur on and within the solum.

Unit 4

Moderate to steep linear slopes (7°-14°), commonly with donga incision, gully erosion, and deep soils, define this geomorphic unit. In areas of great dissection and local relief, unit 4 often dominates the landscape and may be exceptionally long, 1 to 2 km. In such cases it is useful to divide the unit into three equal subunits on the basis of slope length.

Geomorphic unit 4 corresponds to the constant slope segment also termed the Richter slope. There is a balance between the soil material gained from upslope and that lost to the lower units in the landscape. Unit 4 corresponds to the debris slope element of King, the lower backslope of Ruhe and Walker, and landsurface unit six of Conacher and Dalrymple.

Tree growth is variable and depends on the nature and depth of the underlying rock and the structure, fertility status, and texture of the relatively thin colluvial mantle. Weed problems are often found on this unit, in exotic plantations, since there is usually a favorable moisture regime in the topsoil.

Unit 5

Colluvial or alluvial deposition from steeper upslope units and slopes that are concave in profile and vary in angle from 3° to 7° define unit 5. Ephemeral streams typically give this unit a pronounced roughness. In semiarid conditions, this unit is commonly limited to alluvial fans, but in humid climates unit 5 is found at the footslope as a colluvial apron. Unit 5 is of greater lateral extent at the valley head than on side or nose slopes (Young 1975).

The soils show signs of hydromorphy unless the parent material is coarse textured. Deposits usually have incipient stratification, and watercourses are often limited to underground channels that have collapsed in parts. The drainage potential of the underlying materials determines tree growth on this unit. Groundwater iron-pans, plinthite development, and gleying are common soil features. Soils are particularly prone to erosion when sodium and magnesium ions are present in the environment.

The toe slope of Ruhe and Walker, upper pediment element of King, and landsurface unit six of Conacher and Dalrymple are roughly equivalent to geomorphic unit 5. Local convexity due to stream and gully incision is also present in this unit in the drier climates.

Unit 6

Flat surfaces, with slopes of 0° to 5°, and permanent water define this unit. Bottomlands, river terraces, and shallow depressions on extensive plateaus are the landscape features associated with unit 6. Materials in which the soils are developed were deposited or have accumulated as a result of water action.

Soils may vary from well-drained fertile silt loams to relatively deep acid peat deposits. Tree growth is accordingly diverse. In bottomland and stream bank situations, exceptional growth may be present; often these sites are reserved for high-quality veneer or furniture wood production. Flooding is usually a potential hazard, and unit 6 sites that are used for recreation require careful analysis.

This unit comprises the lower toe slope of Ruhe and Walker, lower pediment element of King, and landsurface units seven, eight, and nine of Conacher and Dalrymple.

Most attention has been paid to profile form in the definition of geomorphic units. Horizontal or plan form is irregular, and units will merge one with another in the landscape, primarily as a result of variations in the orientation of gravitational forces.

APPLICATION

Problems of plantation management and silviculture in areas that contain steep topography, complex soil patterns, and varied microclimate can only be solved by an intensification of management (Patterson 1975). The geomorphic approach to site delineation given above will serve as a rough framework or preliminary stratification procedure in cases of exceptional environmental complexity.

The approach advocated here works best in humid and subhumid climates, where relief is strongly expressed and lithology has a dominant influence on landform development. In these steep areas, soil formation is usually minimal, and the major environmental properties of concern to the forester are hydrological regime, stability of the regolith, and structural strength of the soil rather than soil morphology per se.

Covariance between soil properties and topography has not always been found to be satisfactory in denudational terrains (Bleeker and Speight 1978). Sudden soil changes and relatively small homogeneous soil bodies are common features, where the movement of water, minerals, and fabric is lateral rather than vertical within the profile. Geomophic units do not require expensive laboratory analysis, numerous auger observations, or the digging of soil pits. Decisions on site or unit boundaries can be made in the field, without the delay of checking laboratory results needed for pedological surveys.

Because geomorphic units take cognizance of slope and form, sites delineation of this kind is useful for terrain classification, exploitation, planning of extraction methods, deciding on machinery requirements, and determination of road spacing (Rowan 1970). Individual sites must occupy a certain minimum area and correspond with the smallest feasible management unit that can be economically handled.

Geomorphic units proved to be of great value for site index and mean annual increment prediction in a growth study of *Pinus patula*, conducted along the eastern escarpment of South Africa (Grey 1979). Slope percentage, distance from the ridge crest, and shape of the surface have featured consistently in site-factor studies (Carmean 1975).

The efficiency of site surveys can be increased by using easily identifiable features of the landscape. Surface configuration is the one property of ecological significance that is easily perceived, measured, and appreciated in the field. Digital terrain models, computer processing techniques, and methods for the display of three-dimensional surfaces make this approach amendable to automation (Gossard 1978; Gomez and Guzman 1979). Improvements in data capture techniques from remote sensing will also help to reduce the subjectivity involved in mapping geomorphic units (Speight 1977).

A geomorphic approach to site delineation enables the manager of exotic plantations to make a stocktaking of the resources at his disposal, and assists in the evaluation of management opportunities and risks. A common approach will facilitate communication between manager, researcher, and the general public, to their mutual benefit.

REFERENCES

Beckett, P. H. T. 1974. The statistical assessment of resource surveys by remote sensors. In: E. C. Barrett and L. F. Curtis, eds., Environmental remote sensing, 1:11-27. Arnold, London.

Bleeker, P., and J. G. Speight. 1978. Soil-landform relationships at two localities in Papua New Guinea. Geoderma 21:183-198.

Borne, R. 1931. Regional survey and its relation to stock taking of the agricultural and forest resources of the British Empire. Oxford Forestry Memoir 13. Clarendon Press, Oxford.

Carmean, W. H. 1975. Forest site quality evaluation in the United States. Advances in Agronomy 27:209-269.

Conacher, A. J., and J. B. Dalrymple. 1977. The nine unit landsurface model: An approach to pedogenic research. Geoderma 18:1-54.

Gomez, D., and A. Guzman. 1979. Digital model for three-dimensional surface representation. Geo-Processing 1:53-70.

Gossard, T. W. 1978. Applications of TDM in the Forest Service. Photogrammetric Engineering and Remote Sensing 44(12):157-158.

Grey, D. C. 1979. Site quality prediction for *Pinus patula* in the Glengarry Area, Transkei. S.A. For. J. 111:44-48.

__________. 1980. On the concept of site in forestry. S.A. For. J. 113:81-83.

Howard, J. A., and C. W. Mitchell. 1980. Phyto-geomorphic classification of the landscape. Geoforum 11:85-106.

King, L. C. 1967. Morphology of the earth. 2nd ed. Oliver and Boyd, London 726 pp.

Linton, D. L. 1951. The delimitation of morphological regions. In: L. D. Stamp and S. W. Wooldridge, eds., London essays in geography, pp. 199-217. Longman, London.

Litton, R. B. 1972. Aesthetic dimensions of the landscape. In: J. V. Krutilla, ed., Natural environments, pp. 262-291. Published for Resources for the Future Inc., John Hopkins University Press, Baltimore.

Mitchell, C. W. 1973. Terrain evaluation. Longman, London. 221 pp.

Patterson, D. B. 1975. Silvicultural practices on extensive areas of complex sites. Scottish Forestry 29:140-144.

Rowan, A. A. 1970. Terrain classification. Forestry Commission. Forest Record 114. H.M. Stationery Office, London. 22 pp.

Ruhe, R. V., and P. H. Walker. 1968. Hillslope models and soil formation. I. Open systems. Transactions Ninth International Congress of Soil Science, 4:551-560.

Savigear, R. A. G. 1965. A technique of morphological mapping. Annals of the Association of American Geographers 55:514-538.

Speight, J. G. 1977. Towards explicit procedures for mapping natural landscapes. Division of Land Use Research, Technical Memorandum 77/3 CSIRO, Canberra. 4 pp.

Wright, R. L. 1973. An examination of the value of site analysis in field studies in tropical Australia. Zeitschrift für Geomorphologie (suppl.) 17(2):156-184.

Young, A. 1975. Slopes. Longman, London. 288 pp.

3

SYSTEMATIC EVALUATION OF FOREST SOIL QUALITY IN RELATION TO WATER CONSERVATION

K. Takeshita

INTRODUCTION

This chapter approaches the water dynamics of the forest land system through an analysis of the forest environment--that is, the structure of the forest stand, ground surface, soil, topography, geology, and so forth, in the humid-temperate mountain region.

Water delivered to the earth's surface by sporadic rain may be viewed as flows alternating with storages (very slow movement), as the water moves from one zone of the forest land system, which is mainly composed of soil, to another. In each zone, storage affords an opportunity for the redistribution of the water into different outflow channels such as downward and lateral flows, each of which depends on environmental conditions. As a result, the total discharge from the system usually will be steady and continuous.

It is well known that the discharge from a watershed is regulated effectively by environmental conditions. This concept is extended in hydrology to the overall water budget, which is expressed in numerous models and prediction procedures. These models and procedures have proved their worth, but the exact structure of the environmental system is still a mysterious unknown. This chapter strives for greater clarification of the mechanisms of the forest land system and for their evaluation in relation to the conservation of water.

HYDRODYNAMICS OF THE FOREST LAND SYSTEM

Figure 1 shows, in simplified form, the major forms of the storage of water at and near the ground surface of a forest. The water is stored or moves in the environment of plant foliage, ground vegetation, soil bodies consisting of organic matter and weathered rock particles, and rocks that harbor groundwater.

The five principal storage and percolation zones (vegetation canopy, litter cover, ground surface, soil, and bedrock) are stacked more or less with a vertical profile. While gravitational force is always present, not all the forces that move water from one storage zone to another act in a downward direction, but water in all stages moves horizontally over the saturated zones.

The important characteristic of each zone is that it receives a certain input and provides for dividing the output, whose form and variations are changed. Under the ground surface, the several storage zones produce a steady output of water. These main characteristics are explained by the construction of the soil layers and their physical properties.

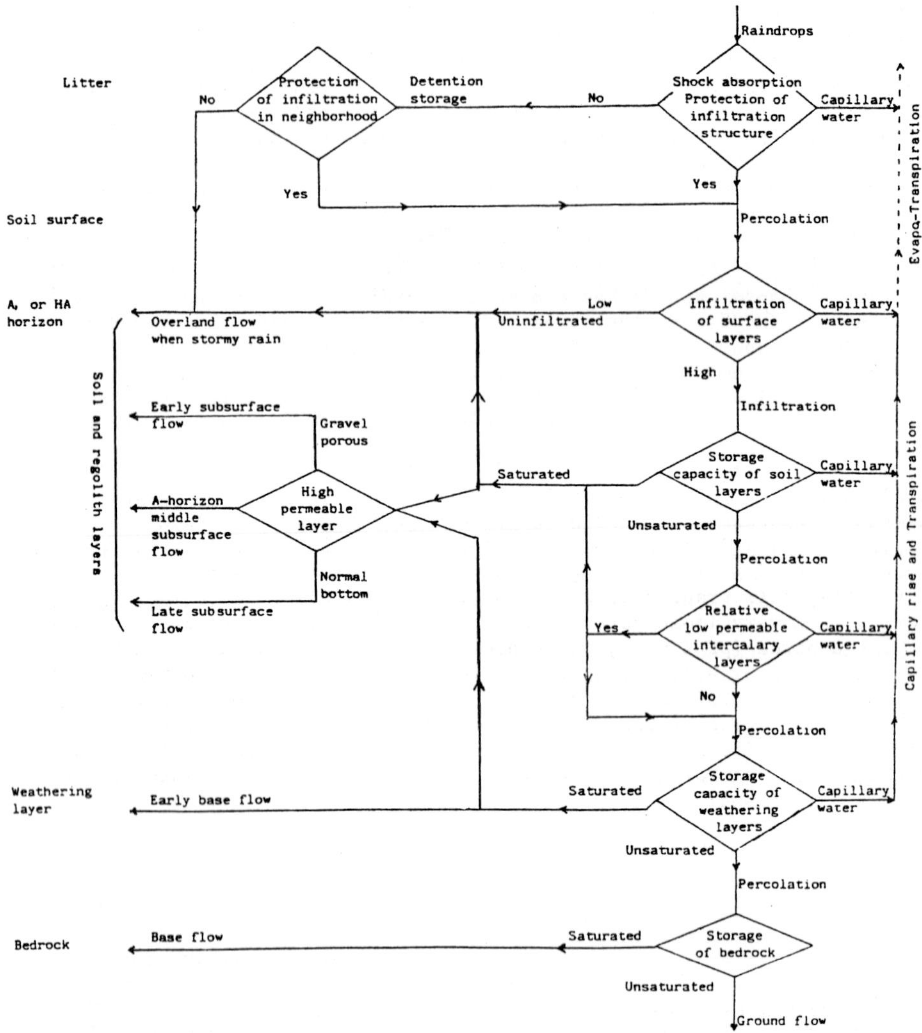

Figure 1. Flow chart of water movements in forest ground.

DISTRIBUTION OF THE TYPES OF DEPOSITIONS ON MOUNTAIN SLOPES

Depositions on mountain slopes are formed by agencies of slope formation and weathering. Debris and particles weathered from rocks are transported downward, and deposits settle at repose angles. In the humid-temperate region, the agencies range from fluids, with different water contents, to solid materials, such as colluvial soil, mud-debris flows, and landslides. For example, a mud-debris flow shows actions of erosion, equilibrium, and deposition depending on the degree of slope and the rate of mixture with water, the matrix (fine-grained material such as clay, silt, and sand), and the debris in the process of moving.

The angle of repose becomes more steep with arid conditions, angular gravel, and root systems of supporting forests. The more that agencies are wet (fluvial) and matrix grains are clayey, the more gentle are the slopes. As a result, distribution of the types of deposits is regulated by both the degree of slope and the slope form (see Figure 2).

SOIL PROFILE AND PORE SYSTEM

Water in the soil is in one of the following three forms: (1) capillary water in fine pores (pF > 2.7), where it remains until extracted by plant roots or evaporation; it moves only by capillary action; (2) gravitational water in loose pores (0.6 < pF < 2.7), where it keeps moving downward; moving by both gravity and capillary action, these pores are of two ranks, loose pores (1.6 < pF < 2.7, with a long storage period) and very loose pores (0.6 < pF < 1.6); (3) gravitative free water in large pores (pF < 0.6), which promptly drains downward and laterally to the side by gravitation; these pores are divided into two ranks, large pores (0 < pF < 0.6) and very large pores (pF < 0). Downward draining systems are formed by root channels, animal holes, soil cracks among gravel in the depositioned strata, huge pipes originating in cracks of rocks and in weathering layers, and large pores among aggregate structure in the A horizons.

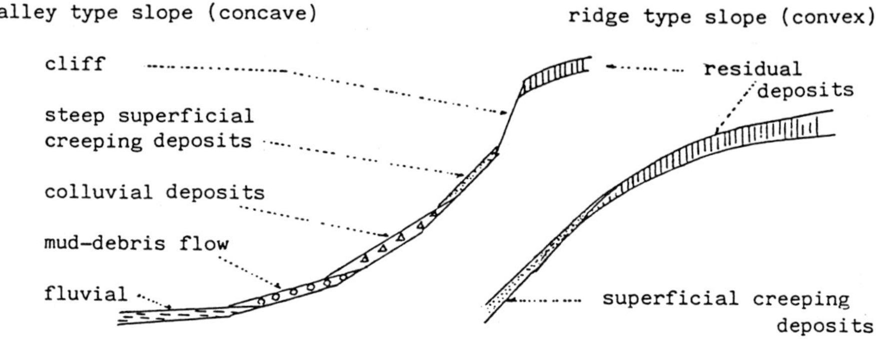

Figure 2. Relation between distribution of types of deposits and topography (Takeshita 1964). Straight-convex steep slope (25°-40°) and straight-concave gentle slope (<25°) are covered with superficial creeping soil.

The occurrence of loose and large pores in soils decreases from the upper to the lower horizons, A_1 (HA) > A > B > C. This tendency to decrease is striking for large pores: there is a large concentration in the A_1 (HA) horizon but little in the C horizon. The humus content in soil has a positive relationship to porosity. The loose and large pores are supported by aggregate structures such as those termed crumby or blocky, whose main structural materials are supplied by humus. In the humid-temperate forest, the decomposition of litter is adequate to produce the aggregates that seem to have the most effective forest influence.

Table 1 shows the occurrence of loose and large pores in different types of deposits. There is a high occurrence of large pores in gravitational deposits and a low occurrence in residual clayey or fluvial-muddy deposits. But the total space of loose and large pores is recognized in thick deposits.

Another characteristic of a soil body as an environment for water is its irregular stratification. Typical soils have clayey horizons 20 to 40 cm under the soil surface, but their actual depth is irregular because of disturbance by erosion. Thus their low porosity is not recognized by average values such as in Table 1. On a mountain, normal deposits are composed of many layers of varying size, porosity, and permeability, but the extent of their area and depth is irregular.

MOVEMENT OF WATER INTO THE SOIL UNDER A FOREST

When rainfall arrives at the soil surface, it is pulled into the pores of the soil by the process of infiltration, but this infiltration sometimes is disturbed by raindrop destruction of the pore structure.

Soil aggregates are reduced to single grains by the kinetic energy of the raindrops, and the infiltration rate of the soil surface is lessened. The magnitude of this kinetic energy depends on the size of the raindrops, their falling speed, and their intensity. The impact of drops of intercepted rain falling from the forest canopy six meters above the soil surface releases as much energy as that of raindrops falling directly from the sky, because the canopy drops are larger (Tsukamoto 1967).

Under the humid-temperate forest, the protection from raindrop impact is provided primarily by the litter layer and secondarily by low vegetation such as grass and shrubs. When water reaches the ground surface faster than it can enter the soil, it ponds into detention storage. The detention storage supplies two major outflows of water: one goes down into the soil (infiltration flow), and one flows off the soil surface (surface runoff; overland flow) (Miller 1977).

Movement downward into the soil is a process characterized mainly by the loose porosity of the layers. The residual water builds up to a stream flow that begins to move laterally, downslope, and away from the site. If the infiltration rate is slow, the detention film deepens and supports the off-site flow; if it is rapid, water enters the soil, and storage and runoff at the surface cease.

On humid-temperate forest land, surface runoff is rare because of the rapid infiltration (> 50 mm/min.) made possible by litter. Although the interception of precipitation by litter and low vegetation has only a small direct function in the redistribution of water, its indirect function for soil conservation is quite remarkable. Similar movements, both downward and laterally, from storage zones sometimes occur in a soil

Table 1. Thickness in different soil horizons of deposits and the occurrence of loose and large pores therein (Takeshita 1977).

Horizon		Dissected Mountain				Undissected Gentle Slope	
		Colluvial	Superficial creeping	Steep creeping	Residual	Superficial creeping	Residual
A_1	Thickness (cm)	5.0	4.7	3.8	2.5	2.2	2.5
(HA)	Large pores (%)	16.7	16.3	19.3	16.4	16.3	16.5
	Loose pores (%)	23.9	24.2	25.3	21.0	24.6	23.6
	Large, loose space (mm)	20.4	19.1	16.9	12.7	9.4	10.0
A	Thickness (cm)	28.2	21.5	16.3	18.0	40.0	24.0
	Large pores (%)	8.4	8.9	12.1	9.3	7.4	9.0
	Loose pores (%)	24.0	23.6	25.8	22.1	20.7	21.7
	Large, loose space (mm)	91.4	69.8	61.7	48.7	106.2	61.8
B	Thickness (cm)	102.4	82.7	82.3	46.0	122.0	100.0
	Large pores (%)	4.8	5.8	6.4	4.2	4.8	4.1
	Loose pores (%)	22.3	22.5	24.3	17.9	18.8	18.7
	Large, loose space (mm)	277.2	234.1	252.7	101.6	288.0	228.0
C	Thickness (cm)	80.5	62.9	34.0	39.0	128.0	135.0
	Large pores (%)	2.0	2.2	2.0	1.6	1.9	1.4
	Loose pores (%)	12.0	13.7	12.4	11.5	11.5	11.5
	Large, loose space (mm)	112.7	100.1	49.0	51.1	163.8	174.2
Total	Thickness (cm)	216.1	171.8	136.4	105.5	292.2	261.5
	Large pore space (mm)	97.0	88.7	86.5	44.0	114.5	79.1
	Loose pore space (mm)	404.7	334.4	293.8	170.1	452.9	394.9
	Large, loose space (mm)	502.0	423.1	380.3	214.1	567.4	474.0

body and weathering layers under it. This is the most important function of storage, providing for a partitioning of water in forest ground.

When water fills all the loose pores on pipe walls of channels or in the soil overlying a less permeable layer, the offside movement of water in the soil (over the intercalary layer or among the aggregate, near the bedrock layer and near the surface layer of A horizon) may become very great. However, the character of water flow in the deep soil seems to resemble that of base flows.

Water percolates downward below the soil layers until it reaches impermeable layers, where it accumulates as groundwater bodies that saturate porous rock layers or unconsolidated materials. Percolated water from the soil is divided into the base flow and the groundwater flow at these storage locations. These water movements in forest land are shown in a simplified flow chart in Figure 1.

THE SOIL ON SLOPES AND ITS CONSERVATION

The magnitude of control of the subsurface flow is characterized mainly by the storage capacities (loose-pore space) of the entire soil body; that is, the subsurface flow decreases and the downward flow increases in accordance with an increase of loose-pore space. The loose-pore space in the entire soil body is calculated from the average porosity and the depth of the soil, but under a normal forest the coefficients of variation of depth are much larger than those of porosity. Therefore, it is considered that in the humid-temperate region, the total loose-pore space of soil is characterized more by depth than by porosity. The depth of soil differs with topography, geology, local erosion, and accumulation.

The porous soil and high infiltration capacity of forest slopes protected by litter permit little rainwater to move over the soil surface. As a result, slopes remain uneroded and the depth of the soil is gradually increased by weathering, aeolian deposits, and subduing agencies.

The stability of soils on steep slopes is supported by the networks of tree roots, but when increased buildup of soil overcomes this stability, landslides result. The interval between landslides on the same site is very long, and is presumed to be several hundred years in the humid-temperate region. Forests support soils of considerable depth during these periods, and the soils provide water conservation.

MODELING OF THE FOREST GROUND SYSTEM

The movement of water in the forest ground system is characterized by many processes, such as infiltration, storage, downward flow, off-site flow (surface flow, subsurface flow, and ground flow), capillary flow, downward permeability, and draining permeability (lateral and horizontal permeability). This activity is simplified in a tank model in Figure 3.

The soil body is expressed as an enclosed tank which is divided into four compartments for convenience. The upper compartment is mainly characterized by litter and A_1 (HA) horizons. The other compartments have three structures: surface layer, middle (inner) layer or soil aggregate, and bottom layer.

Generally, infiltrated soil water fills in fine pores earlier than large pores because of the stronger capillary power of fine pores.

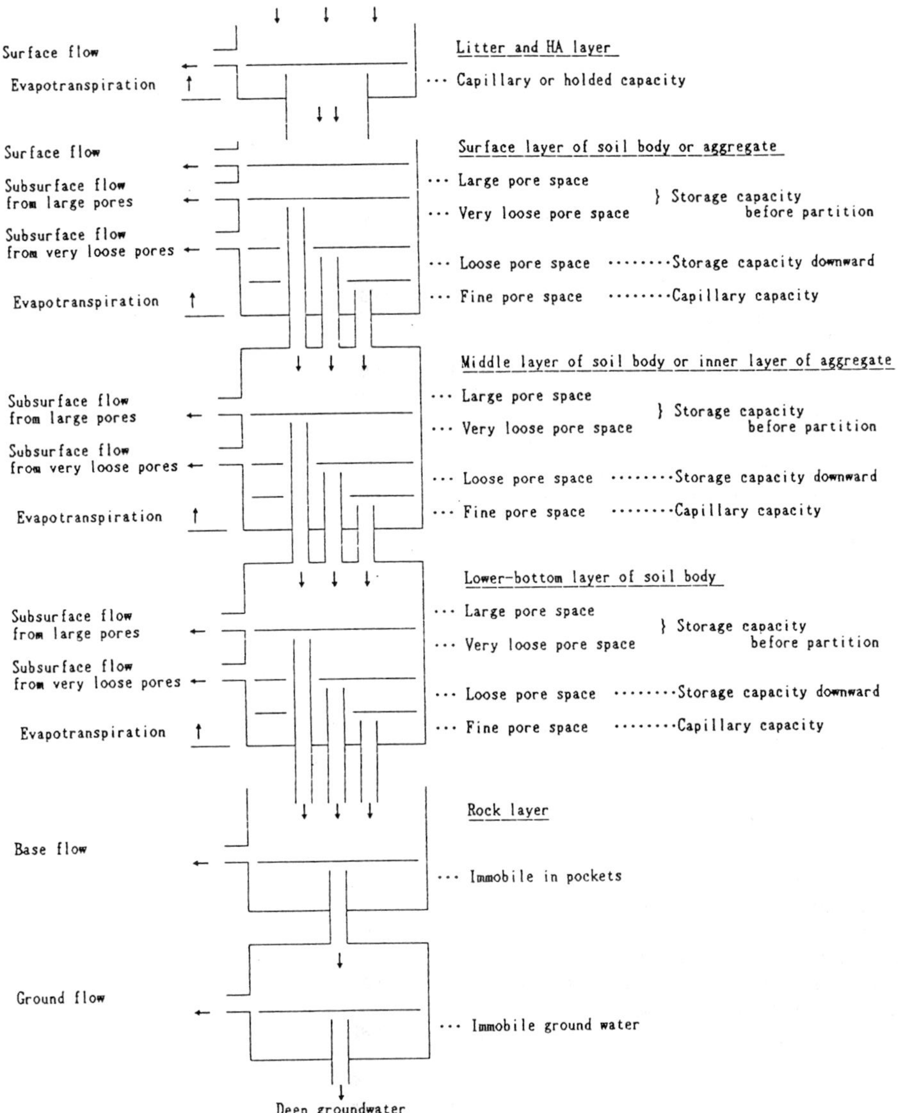

Figure 3. Forest ground tank model.

Therefore, in the compartments of this model, pore spaces (partial capacities) are arranged from bottom to top in the following order: fine pore, loose pore, very loose pore, and large pore. Capacities of each compartment are indicated by the sum of the four kinds of spaces. But, in humid-temperate regions, fine pores and half of the loose pores are saturated by water in normal seasons, so that effective capacities of each compartment are indicated by the large pores, very loose pores, and half of the loose pores except during abnormally arid seasons.

Each compartment has some permeable or draining pipes for downward and lateral movements. Diameters of pipes are characterized both by the size of pores (loose, very loose, and large) and the topography (slope length, slope inclination). Pipes are situated in the upper level of the three spaces (fine, loose, and very loose). Pipes for sideflows are not characterized by loose pores, but by both very loose pores and large pores.

CONTROL EFFECT OF FOREST GROUND ON DRAINAGE DISCHARGE

The base flow of groundwater from deep bedrock is characterized by steadiness. The individual inputs of the percolates from stormy rains are slowed down and stabilized. On the other hand, the subsurface flow from soil layers temporarily saturated with water is characterized by instability. The former is connected with low discharge and the latter with direct discharge (flood flow).

Figure 4 indicates the schematic duration curve of river discharge and precipitation in a year. When daily discharge is nearly the same throughout the year, the discharge-duration curve will be shown by the average line EF, so that this line is considered an ideal controlled curve.

Under part of the average line, rectangle OEFg is equal to the quantity of annual discharge. This is divided into three parts: (1) OEHb is discharge controlled by rainfall itself and short-time storage of soil; (2) bHIdg, patterned by vertical stripes, is discharge controlled by storage of soil and bedrock; (3) IFd, patterned by slanting stripes, is noncontrolled value. If we wanted to get the same level of discharge throughout the year, we would have to build artificial controlling equipment, such as a dam.

It is presumed that the proportion of noncontrolled value is a function of storage capacity, weathering, and topography. For example, the ratios of noncontrolled value to annual precipitation (2,200 ∿ 2,500 mm) in different basins covered with different soil are indicated as follows:

(a) steep mountain covered with large storage capacity soil........ 0.3
(b) volcanic plateau covered with large storage capacity soil...... 0.2
(c) steep mountain covered with moderate storage capacity soil..... 0.4
(d), (e) steep mountain covered with small storage capacity soil (coarse gravel or low infiltration) ....0.5 ∿ 0.6

As an index of uniformities of base flow during a year, the author adopted the difference between a plentiful discharge and a scanty discharge (a specific discharge), and considered that the more this difference decreases the more the degree of uniformity, if the base flow is high. Figure 5 (1) shows the relationships of "plentiful-scanty" discharges, topography, and geology in the case of an annual precipitation of 2,500 mm, where the topography is expressed as a percentage of the area occupied by gentle slopes (<25°). As a result,

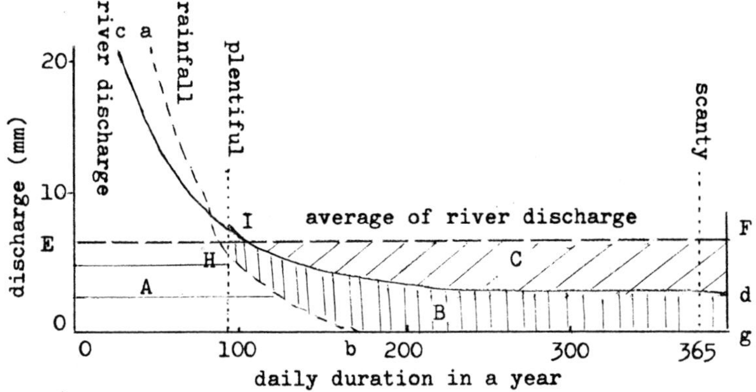

Figure 4. Schematic discharge-duration curve of river discharge and precipitation. A = controlled by rainfall itself and short-time storage of soil. B = controlled by storage of soil and bedrock. C = noncontrolled value.

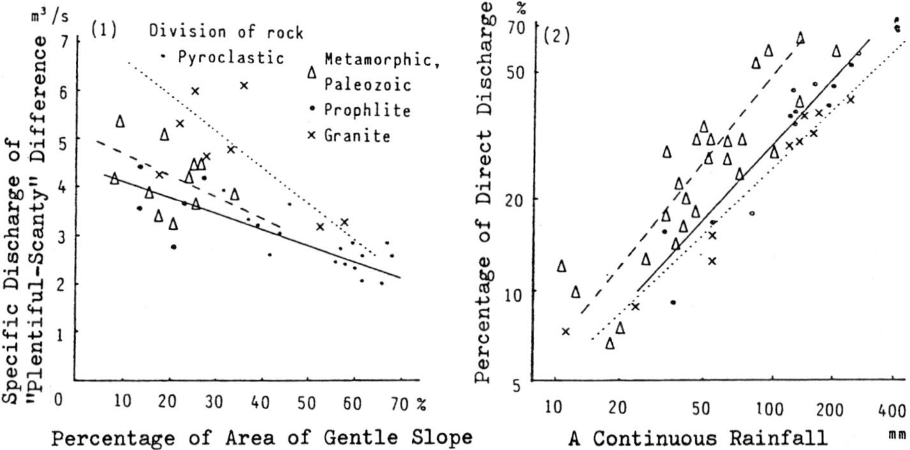

Figure 5. Relations of discharge, topography, and geology.
(1) Relations of specific discharge and "plentiful-scanty" difference, percentage of area of gentle slope, and geology.
(2) Effect of precipitation and direct discharge on different geological formations.

the uniformity is regulated by the areal percentage of gentle slopes, namely the depth of the soil, and the geological characteristics, such as a crushed or deeply weathered zone.

Figure 5 (2) shows the relationships of the direct discharge percentage, the amount of precipitation, and the geological characteristics, under conditions of a specific base flow of 6-15 cubic meters per second on a dissected mountain.

It seems that coefficients of subsurface flow are higher in gravelly soils, such as andesite or schist, than in sandy-muddy soils, such as deeply weathered granite. In weathered granite basins, subsurface flows are relatively small, but early base flows are large. Therefore, the uniformity of discharge is not always high.

The relation between peak value of river discharge and rainfall is considered important in flood control. Recently, however, it has been found that this relation is modified by the nature of soil-weathering layers in the basin. Figure 6 shows different curves for different soil and weathering layers as identified below:

(a) Mountain basin covered with deep, porous soil and weathering rock (high storage capacity) (soil depth >1.5 m)
(b) Gentle volcanic plateau covered with deep, porous volcanic ash and volcanic breccia (high storage capacity) (soil depth >2 m)
(c) Granite mountain basin covered with relatively little soil (<1 m)
(d) Mountain basin covered with very coarse gravelly soil whose loose-pore capacity is small (low storage capacity)
(e) Mountain basin covered with low infiltration soil under subtropical forests (low infiltration)

It is clear that the peak of discharge of high storage capacity basins is smaller than that of low storage or low infiltration in spite of similar rainfall volume. That means that the function of flood control of the former is higher than that of the latter.

EVALUATION OF FOREST SOILS FOR WATER CONSERVATION

In humid-temperate forest ground, influences on the movement of water are characterized by a high infiltration rate at the surface, the presence of loose and large pores (storage capacities), and deepness of soil, all of whose values are large enough to accommodate heavy rains.

It is usually preferable when the discharges from watersheds are uniform, steady, and high all year round, and flood discharges are low. These desirable features correspond to increased occurrence of loose pores and greater depth of soil. The former is accomplished by the decomposition of litter and the intermixture of humus into the soil, with the soil being supported by the root systems of huge forest trees.

Thus the uniformity of discharge is controlled directly by the pore spaces of the soil and the rock layers, but this mechanism does not begin until water infiltrates. Therefore, the high infiltration rate of the surface layer, protected by litter, is the most important role aspect of the process.

The covering ability of litter results from the balance between the falling leaves supplied by the forest canopy and their decomposition to humus. Rapid and perfect decomposition accompanied by heavy surface erosion, such as in tropical areas, results in little soil depth. The stability of the litter supply on steep slopes is supported by the stems of low vegetation, such as grasses and herbs, under the forest. The control of the densities of forest stands, which regulate the intensity of light for grass growing, is important. It is considered that the influences of the covering ability of litter and its decomposition are recognized not only in soil formation and water conservation but also in landscape evolution.

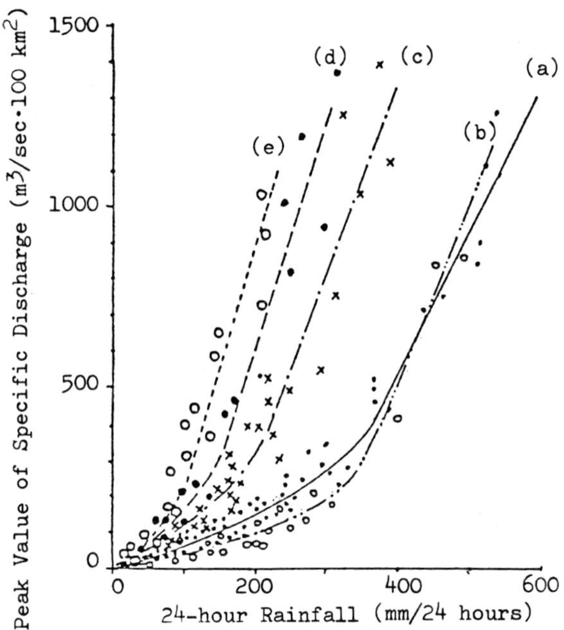

Figure 6. Relation between peak value of river discharge and 24-hour rainfall in different mountain basins having different storage of soil and weathering layers:

(a) high storage capacity soil and weathering layers
(b) gentle slope covered with high storage capacity soil and weathering layers
(c) moderate storage capacity soil
(d) low storage capacity soil
(e) low infiltration soil

REFERENCES

Miller, D. H. 1977. Water at the surface of the earth. Academic Press, New York. 557 pp.

Takeshita, K. 1964. The formation of mountain slope and its meaning to forestry. Fukuoka Forest Exp. Sta. Bull. 17. 109 pp.

__________. 1982. Relation between construction of forest soil and flood discharge. Data analysis natural disaster, No. 9, pp. 72-79.

Takeshita, K., and J. Takagi. 1977. Soil physical and topographical study of water conservation on humid warm temperate forest land. Fukuoka Forest Exp. Sta. Bull. 26. 51 pp.

Tsukamoto, Y. 1967. Raindrops under forest canopies and splash erosion. Tokyo University Agr. and Tech. Forest Bull. 5, pp. 65-77.

4

A SITE CLASSIFICATION FOR FOREST LAND USE IN JAPAN

Y. Mashimo and K. Arimitsu

INTRODUCTION

Forests produce not only such products as logs, pulpwood, and poles but also a variety of functions that contribute greatly to environmental conservation and to upgrading the lives of the inhabitants in a forested drainage basin. Above all, it is well known that the forests function to prevent landslides and washouts as well as to mitigate flood damage in downstream areas. The disaster-prevention function of forests plays a vital role in mountain areas with steep topography, geological weaknesses, and much rainfall.

In Japan, the establishment and consolidation of protection forests have been promoted in mountainous areas that frequently suffer from disasters. A study of recent disaster records shows that it is almost impossible to forecast the time and place that heavy rainfall will occur. Under these circumstances, conditions of topography, geology, and soil structure that influence the occurrence and extent of disasters have been taken as parameters for evaluating the degree of hazardousness of forested mountain land under heavy rainfall. In other words, a graduated rating is given to disaster prevention by a system of territorial classification. Other forestry functions, such as water-resource cultivation, can be classified in terms of their value in the same manner. Integrated evaluation of these functions allows a successful site classification for the proper scheduling of forestry work such as reforestation and other operations on forest lands.

The forestry functions that benefit from site classification are (1) timber production, (2) water-resource cultivation, (3) flood control, (4) soil conservation (erosion control) and landslide prevention, and (5) forest recreation. Increased pressures on forest land are expected, especially as a result of overpopulation in urban areas and the consequent growing tendency of city dwellers to seek out forest land for relaxation. Such activities are specified in Japan as forest recreation.

METHODOLOGY OF SITE CLASSIFICATION OF FOREST LAND

The grid method is often used to evaluate large areas of forest land, and various grid sizes are used depending on the scale of an objective area and the accuracy required for evaluation. Each gridded area is classified in terms of functions as high, medium, and low grade for rating the degree of hazardousness of the forest land, based on the information obtained from topographic and soil maps. The criteria for site classification by forest function are described below.

Timber Production

The current growing stock and tree species are the major parameters for a short-term evaluation of forest lands. However, logging transportation conditions provide another effective parameter. For a long-term evaluation for timber production, the potential productivity of the land to grow wood should be used.

It is considered reasonable to rate forest-land productivity in three grades in Japan: (1) high-productivity land: favorable growth can be expected by reforesting with useful species; (2) medium-productivity land: growth is less favorable than on high-productivity land, although the land is suitable for reforesting with useful species; (3) low-productivity land: the land is not suitable for reforesting with useful species or an average growth of 5 cubic meters per hectare per year cannot be expected.

The productivity of forest land (site quality), which depends on climate and land conditions, can be evaluated by soil type and elevation (temperature index). In Japan, the estimation of site class (40-year-old tree height in meters) by a combination of environmental and soil factors has allowed estimations of productivity by region and tree species with considerable accuracy. A simplified classification standard based on soil type (Forest Soil Division 1976) and elevation is introduced here.

Soil productivity can be rated as follows: (1) high-productivity soils: moderately moist brown forest soils, slightly wetted brown forest soils, and moderately moist black soils; (2) medium-productivity soils: dry brown forest soils and slightly dry black soils; (3) low-productivity soils: regosols, lithosols, podzols, red soils, yellow soils, peaty soils, and gley soils.

Altitudes (as in the mountain area of the Tone River Basin) may be grouped as follows: (1) 0 to 1,000 meters: the soils of the high-productivity group have high productivity, those of the medium-productivity group have medium productivity, and those of the low-productivity group have low productivity; (2) 1,000 to 1,400 meters: most soils in this elevation range have medium productivity, except for those of the low-productivity group; (3) over 1,400 meters: all soils at these elevations have low productivity.

Water-resource Cultivation

A desirable feature of water catchments in Japan is to have precipitation in the headwaters of a basin percolate into the soil. This results in small seasonal fluctuations in the river discharge--specifically providing adequate discharge during the drought season. We consider recharging of the water-resource function to involve four factors: soils, surface geology, slope, and altitude.

Porous, deep soils contribute to the percolation and temporary retention of water. According to Mashimo (1974), forest-land soil in Japan holds an average of about 180 mm of effective water per meter. This is a larger water-holding capacity than for the soils of pasture, farmlands, and housing lots. The soil type and parent material (geology) are dominant factors affecting the water-holding capacity of soils. Soils can be grouped by their water-holding capacity: (1) large water-holding capacity: moderately moist black soils, moderately moist brown forest soils, and slightly wetted brown forest soils; (2) medium water-holding capacity: slightly dry black soils, slightly dry brown forest soils, slightly wetted black soils, slightly wetted brown forest soils, and wet humus podzols; (3) small water-holding capacity: immature soils, dry brown forest soils, red and yellow soils, dry podzols, wet iron podzols, and gley soils.

The bedrock underlying the soil layers has a strong effect on the formation of an aquifer and the storage and movement of groundwater. Weathered granite and volcanic ash, for example, permit a great deal of water to percolate easily. Based on existing data, the geological contribution to the water-resource function can be classed as follows: (1) large pyroclastic materials, volcanic ash, weathered granite, and a fractured zone; (2) medium Quaternary-stratum sand and mud, Tertiary-stratum conglomerate and sandstone, tuff, Mesozoic/Paleozoic sandstone, phyllite, crystallines, and andesite; (3) small Quaternary mud, Tertiary mudstone, Mesozoic/Paleozoic chert, rhyolite, and serpentine.

Percolation of precipitation into the soil is related to slope. A steep slope may suffer rapid surface runoff, and little opportunity for percolation. In Japan slopes are classified as gentle (less than 10 degrees), medium (10 to 30 degrees), and steep (over 30 degrees). Their contribution to water resource cultivation decreases in the order of gentle > medium > steep.

High-altitude sites have lower temperatures than low-altitude sites, and more precipitation. They thus provide greater potential energy of the water retained. In other words, the low-altitude areas are less important than high-altitude areas as water sources. Medium-altitude mountain sites are also important sources. The contribution of altitude to water storage decreases in the order of high > medium > low.

Each of the four factors of soil, geology, slope, and altitude for forest lands has been classified into three grades for evaluation (Table 1). The results expressed in points are totaled so as to have three ratings (high, medium, and low) for the forest land concerned (Table 2).

Table 1. Forest-land evaluation factors and points given regarding water-resource cultivation.

Factor	Points 3	2	1
Soil	Large water capacity	Medium	Small
Geology	Large contribution	Medium	Small
Slope	Gentle slope	Medium	Steep
Altitude		High/medium	Low

Table 2. Rating of evaluation points given regarding water-resource cultivation.

High	Medium	Low
9-11	7-8	4-6

Table 3. Evaluation factors and points given for priority in flood control.

Factor	Points		
	3	2	1
Altitude	High	Medium	Low
Slope	Over 30°	10°–30°	Below 10°
Ravine density		High	Low

Table 4. Rating of evaluation points regarding flood control.

High	Medium	Low
8, 7	6, 5	4, 3

Flood Control

Precipitation, if discharged directly into rivers as surface runoff from forested watersheds, will increase the hazard of flooding in downstream areas. Flood control is crucial in such areas. Altitude, slope, and ravine density are used in Japan to evaluate runoff. Although soil conditions do affect surface runoff, the other three factors are considered to be more important.

In Japan, a drainage basin at high altitudes discharges a relatively large amount of water, as determined from past surveys. Altitude, when classified as high, medium, and low, decreases in the order of high > medium > low in terms of rating for flood control.

The velocity and amount of surface runoff is influenced by the slope of the land, and is greater as the slope becomes steeper. The affect of slope on runoff is in the order of gentle < medium < steep.

Ravine density is used as a parameter in the development of surface runoff courses. Land with a high ravine density allows precipitation to be discharged into rivers and reach downstream areas more quickly. But it is not considered as influential on flood discharge as the factors of altitude and slope, and therefore is classed only as high density and low density.

Evaluation points are given to the three grades of altitude and slope and to the two grades of ravine density (Table 3). The evaluation points for these three factors are totaled so as to have three ratings (high, medium, and low) for the forest land for flood control (Table 4).

Soil Conservation

Soil conservation in mountain areas is conducted mainly to prevent landslides and soil washouts, which are caused mostly by heavy rainfall. But earthquakes should also be counted as a major reason for mountain

Table 5. Evaluation factors and points for soil conservation.

Factor	Points		
	3	2	1
Surface geology	Large hazardousness	Medium	Low
Slope	Over 30°	20°-30°	Less than 20°
Ravine density		High	Low

Table 6. Rating of evaluation points regarding soil conservation.

High	Medium	Low
8, 7	6, 5	4, 3

disasters in Japan. Slope, surface geology, and ravine density are adopted as the evaluation factors. Some soils are more vulnerable to landslides and erosion, and need special treatment, but soil classification for this purpose has not been developed in Japan as yet.

Landslides rarely occur on slopes under 20 degrees, but in limited areas (average 0.2 ha) they occur frequently on steep slopes and cause slippage. This makes it difficult to judge the extent of hazardousness from the average land slopes framed on the grids. But generally it can be said that an area with steep slopes will have many steep slopes. The relationship between slope and the decreasing frequency of landslides is given in this order: gentle (less than 20°) < medium (20°-30°) < steep (over 30°).

Several analyses on the causes of landslides have been attempted based on many past landslides, and surface geology is considered as one of the important factors for such analyses. Surface geology can be categorized according to the possibility of causing landslides as follows: (1) large hazard: pyroclastic materials, weathered granite, unconsolidated deposits; (2) medium hazard: late Tertiary deposited rock, agglomerate, rhyolite, granite, black schist; (3) small hazard: Mesozoic/Paleozoic deposited rock, andesite, basalt, green schist.

Areas with high ravine density, providing more ravine heads and faces, are prone to have more landslides, although ravine density is not as influential as slope and geology. The classification of ravine density is "high" and "low."

Evaluation points are given to three grades of slope and surface geology and to two degrees of ravine density in determining soil conservation (Table 5). Classification is made regarding the rating of forest land based on the total points for the three factors (Table 6).

Forest Recreation

People living in densely populated urban areas have become eager to relax through recreation in forests and fields, rivers and lakes.

Increasing income and leisure time have expanded the demand for outdoor recreation. Forests can provide the facilities to meet such increasing requirements of city dwellers.

Forests provide these people with (1) scenic beauty, (2) campsites, (3) sport fields, and (4) trails for climbing. In addition, Japanese forests have many interesting things to see, such as historical remains, religious objects, and academic monuments. These valuable things, however, are apt to be appreciated only subjectively, from metaphysical and aesthetic viewpoints. In order to provide an objective evaluation method, the following criteria have been established: (1) high value: specially designated national parks and quasi-national parks; specially designated areas reserved for wildlife; specially designated areas for conservation of the natural environment; special recreation forests, scenic beauty protection forests, and prefectural recreation forests; important forests reserved for recreational activities, areas commanding scenic views, and areas adjacent to lodging facilities; and historical monuments, tourist sites, and natural monuments designated by the national and local governments; (2) medium value: ordinary national and quasi-national parks; ordinary areas reserved for wildlife; and areas within five kilometers of important sites commanding scenic views; (3) low value: forest land lacking the above designations.

REVIEW OF THE RESULTS OF EVALUATING FOREST LAND

The estimation of timber production based on site-quality parameters and developed by a quantified evaluation method is regarded as having a high degree of accuracy. However, in studying the correctness and consistency of the results, explicit standards should be used, such as the growth status of reforested stands, reforestation ratio, major tree species in natural forests, and actual results of past production.

In the study of water-resource cultivation and flood control, recorded observations of river discharges will allow comparison of the characteristic features of segmented drainage basins. For the cultivation of water resources the base-flow discharges serve as parameters, whereas for flood control the flood discharges serve as parameters. For instance, a comparison between segmented drainage basins of maximum specific flood discharges of the past thirty years can be used as a test. If the segmented drainage basins with large proportions of high-rating grids for flood control have larger specific discharges than those with small proportions, then results are consistent. The availability and size of dams and water utilization in the basin also affect flood control. The extent and nature of forests in the basin may also affect the discharge of the river.

In studying soil conservation, the frequency of past landslides and the required expenditures for rehabilitation works can serve to review the accuracy and consistency of the evaluation. Furthermore, sediment in a reservoir will be useful as an object of review, as well as the existence of various disaster-protection forests, landslide sites, and soil and water trapping devices.

In regard to forest recreation, the amount of tourist influx into forest areas and the amount of money spent therein are important references for reviewing the evaluation ratings.

SITE CLASSIFICATION BY PRIORITY OF FOREST LAND AND PROPOSED LAND USE (CASE STUDY OF A MOUNTAIN AREA IN THE TONE RIVER BASIN, JAPAN)

This section, taking the Tone River Basin as an example, presents a case study of site classification of a mountainous area, the desirable arrangement of forest land, and related work.

The Tone River runs through the center of the Kanto Plain, supplying it with water. Tokyo receives more than half its large water supply from the river. The catchment area, the largest in Japan, is about 16,000 square kilometers. The Tone is considered the most important river in Japan. The headwaters, in the northwestern mountainous area of the Kanto Region, have an area of 6,200 square kilometers covered with forest.

This objective area was covered with a grid of 4 square kilometer squares. There were 1,561 grid squares, each of which was examined for its average slope, altitude, and ravine density. Furthermore, much information was interpreted from related soil and geological maps to make the classification ratings by function.

The five classification grid maps (Figures 1, 2, 3, 4, and 5), each containing a range of ratings, were superimposed over each other to reveal areas with specific features. Then the mountainous area of the basin was divided into sixteen blocks involving nine mountain areas, five volcanic areas, and two hilly areas (Figure 6). Natural geography and administration divisions were also taken into account in this deliniation. The study of the desirable forest land use for these resulted in the four patterns discussed below.

1. Areas aimed at water conservation and forest recreation. Forest land classified as having "low" timber productivity but high ratings in other functions should be so designated, to make it possible for the forest to provide for effective public use. In particular, areas K, L, and M (Figure 6 and Table 7) include Nikko and Oze, which are famous in Japan for their scenic beauty. These areas are important for their recreational and tourist sites and also can be given high ratings in flood control. Since they are very poor in tree growth, the existing forest should be reserved and upgraded as soon as possible by actions that include prohibition of cutting.

Areas of A, B, and E provide some recreational sites, although not as important as those mentioned above, and therefore should be designated for water conservation. These areas have comparatively "low" productivity as forest land, and none of them are suited for timber production.

2. Areas aimed at timber production together with water-resource cultivation and its improvement. In areas H, with "high" timber productivity, and D, with comparatively "high" timber productivity, useful species can be planted, but the cutting age should be raised and the cutting area reduced. Even newly reforested areas can function successfully in cultivating the water resource if stocking is abundant and crown coverage is high.

In areas C, F, and G, which provide "medium" productivity, site selection should be made prudently from the standpoint of soil and topography for successful reforestation. Cutting age and areas should be determined with restraint as mentioned above.

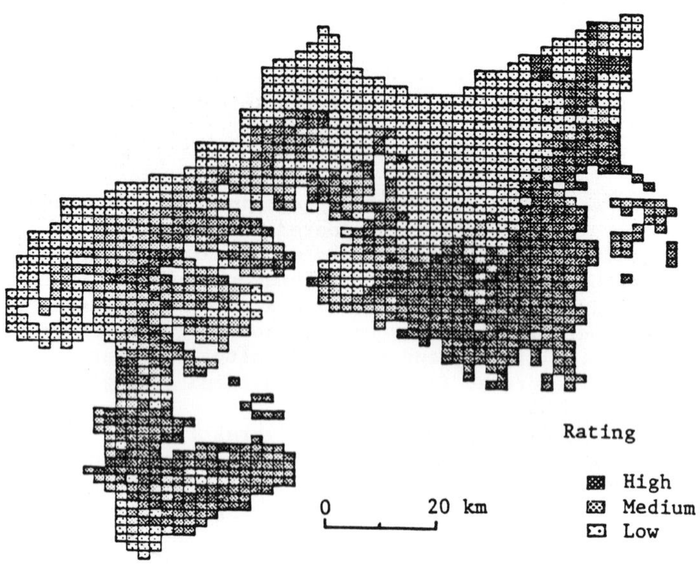

Figure 1. Grid map rating the function of timber production.

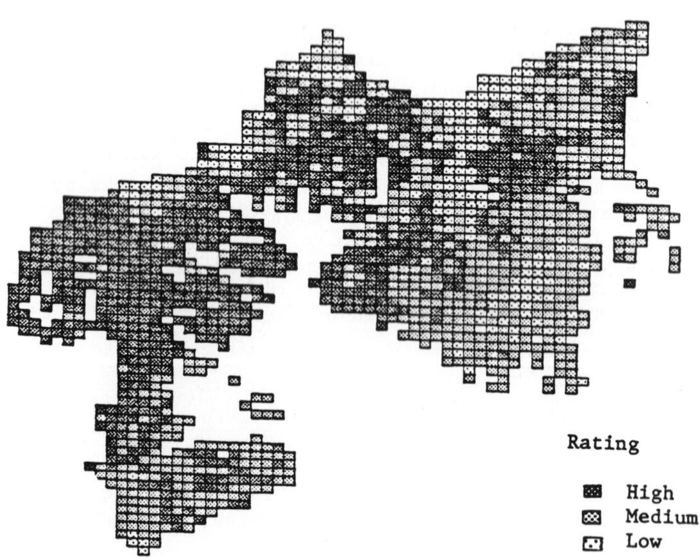

Figure 2. Grid map rating the function of water-resource cultivation.

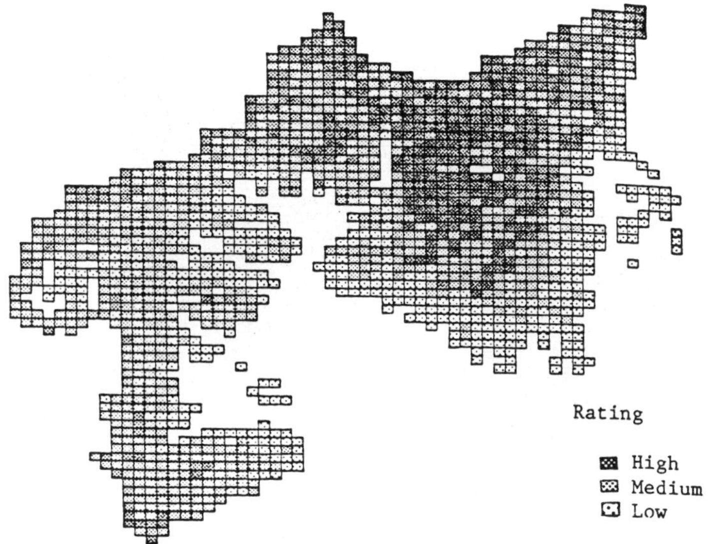

Figure 3. Grid map rating the function of flood control.

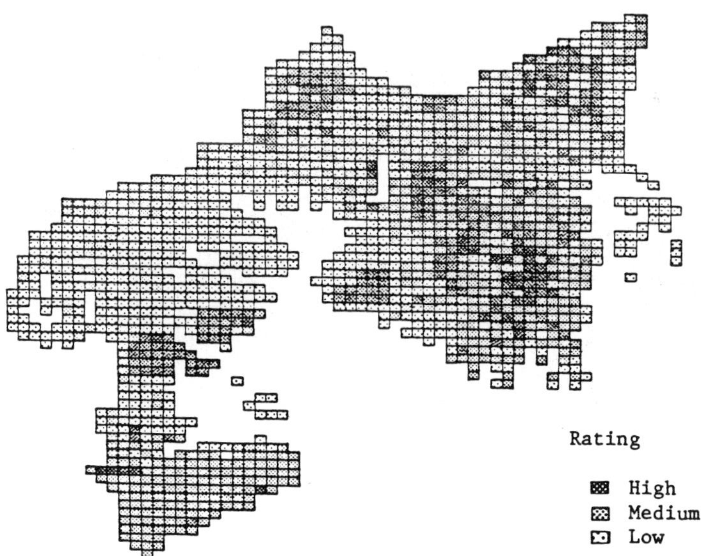

Figure 4. Grid map rating the function of soil conservation.

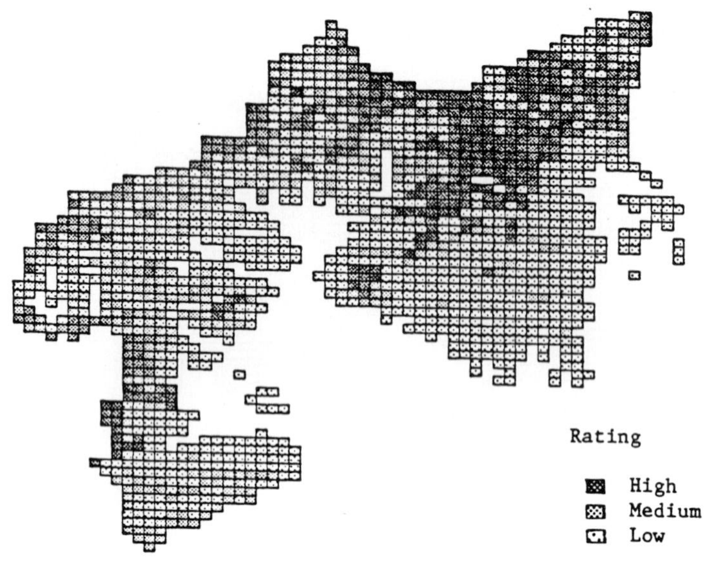

Figure 5. Grid map rating the function for recreation site.

Figure 6. Divided sixteen blocks in the Tone River basin.

Table 7. Evaluation of functions in grids of sixteen blocks in the Tone River Basin.

	Block:	A	B	C	D	E	F	G	H	I	J	K	L	M	N	O	P	(Total)
	Grids in Number:	119	102	114	127	51	75	57	117	138	7	133	22	136	61	276	26	1,561
	%:	7.6	6.5	7.3	8.1	3.2	4.8	3.6	7.5	8.8	0.4	8.5	1.4	8.7	3.9	17.6	1.6	(100)
Function	Rating																	
Timber production	High	4	0	50	58	7	22	17	70	99	7	15	7	5	50	254	24	689
	Medium	6	16	58	44	3	40	24	36	23	0	5	2	1	4	8	2	272
	Low	109	86	6	25	41	13	16	11	16	0	113	13	130	7	14	0	600
Water-resource cultivation	High	39	98	108	77	40	73	47	85	41	0	33	16	77	18	25	6	783
	Medium	39	4	5	45	8	2	10	32	93	7	57	6	25	41	207	20	601
	Low	41	0	1	5	3	0	0	0	4	0	43	0	34	2	44	0	177
Flood control	High	64	12	0	25	25	2	0	4	16	0	102	7	112	22	49	0	440
	Medium	55	75	76	89	26	35	22	87	93	0	31	12	23	23	91	0	738
	Low	0	15	38	13	0	38	35	26	29	7	0	3	1	16	136	26	383
Soil conservation	High	28	0	0	12	3	16	16	30	6	0	41	5	38	14	60	0	269
	Medium	63	17	11	57	19	5	4	26	127	1	82	5	76	34	156	0	686
	Low	28	85	103	58	29	54	37	61	5	6	10	12	22	13	60	26	609
Forest recreation	High	48	32	7	16	12	5	9	38	1	0	73	19	99	23	7	0	398
	Medium	63	40	55	52	18	25	13	29	49	0	28	0	15	3	16	0	406
	Low	8	30	52	59	21	45	35	50	88	7	32	3	22	35	253	26	757

3. Areas aimed mainly at timber production. Areas of N, I, and O in Figure 6 have forest land with promising timber production, whereas they are "medium" or "low" in other public functions. Forestry in these areas should encourage timber production, particularly in parts of I and O that have "very high" land productivity for Japan. Intensive forestry operations are expected in these areas as the completion of the infrastructure is advanced.

The mountains in these areas have topographic conditions in the mature stage, and slopes of about 30 degrees are developing. Although these slopes are comparatively stable, they have a "medium" degree of potential danger for landslides. Therefore, great care should be taken with local steep slopes to prevent land slippage.

4. Areas aimed at conservation of the environment. Forests on hilly land close to urban areas are highly valued for disaster prevention and scenic beauty. Recently, however, some of this valuable forest land has been destroyed by the unorderly changes in land use and the large-scale changes of topography. Forest destruction naturally causes many problems for disaster prevention, but a much more serious problem is the loss of a comfortable living environment. No compensation will be paid for a high-quality living environment once it is lost, even with the efforts of complete disaster-prevention facilities provided by the cities and towns. The forest land in areas J and P adjacent to Takasaki and Utsunomiya, in Gumma and Tochigi prefectures respectively, should be guarded in this respect.

APPLICATION

This approach to the evaluation of the functions of forest land, site classification, and the application of grids to objective areas is used for forest land-use planning in Japan. Particularly, the imposing of a grid on a study area is a necessary procedure in making a forest plan.

The privately owned forests in Japan are divided into 256 districts for forest planning, whereas the national forests have 80 working blocks throughout the country. For each district and working block, a ten-year forest plan is made every five years, and this plan requires a "forest function survey" to evaluate the forest's potential for timber production, water-resource cultivation, disaster prevention, and recreational sites. The survey results are used not only by the forestry administration in making forest plans and their implementation but also for administrative work in other fields related to forestry and for management by forest industries.

Although the grid method for evaluating forest land by function is being used throughout Japan, it is still open for consideration which factors should be taken up for each function, what parameters should be applied for evaluation, and how to make the classification of ratings of hazardousness. Further studies are also needed on a variety of factors, such as soils, vegetation, geology, topography, climate, and the cause-and-effect relations among them, in order to establish more accurate grids and classification methods.

REFERENCES

Forest Soil Division. 1976. Classification of forest soil in Japan. Gov. For. Exp. Sta. Bull. 280.

Mashimo, Y. 1974. Estimation of stand growth by quantification of soil conditions and environmental factors. Transactions Tenth International Congress of Science, 6(1):50-55.

RUNOFF RESPONSE TO PEATLAND FOREST DRAINAGE IN FINLAND: A SYNTHESIS

M. R. Starr and J. Päivänen

INTRODUCTION

The influence of forest drainage on peatland hydrology is not without controversy in Finland. Both beneficial and detrimental changes in runoff following drainage have been reported. Because of the large area of peatland that has been drained for forestry purposes, the debate is particularly important. Of approximately 9.7 million hectares of the land area covered by peat, perhaps half has been drained.

Forest drainage of peatlands has been mainly criticized because it is considered to have increased the frequency of flooding downstream. Since the occurrence of flooding is related directly to the size of the peak flows, the essential question is, does drainage increase peak flows?

The aim of this chapter is to review the results of several recent investigations dealing with peatland forest drainage and peak flows. Emphasis is laid upon the runoff leaving the actual ditching area (e.g., Päivänen 1974; Heikurainen 1976 and 1980; and Ahti 1980), rather than whole catchments (e.g., Mustonen and Seuna 1971; Seuna 1974 and 1980) or river basins (e.g., Hyvärinen and Vehviläinen 1978).

HYDROGRAPH MODELS

Runoff response can be described graphically by the "hydrograph," the characteristics of which are given in Figure 1. Comparisons of

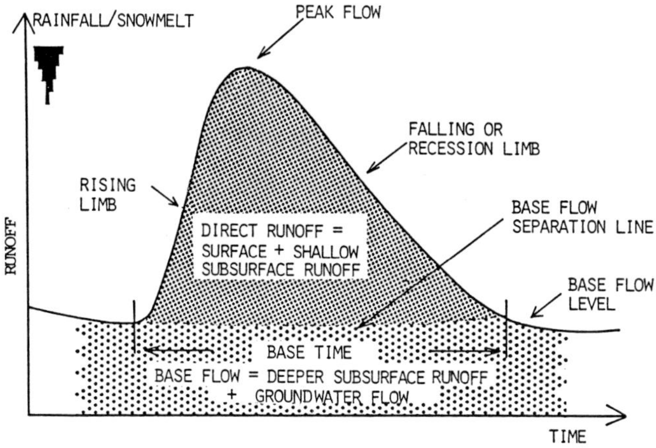

Figure 1. The characteristics of a peak flow hydrograph.

hydrographs from drained and nondrained sample catchments in Finland by different workers, however, produce conflicting results. A similar situation has been noted in the results from the Federal Republic of Germany (Baden and Eggelsmann 1968) and from the United Kingdom (Green 1973). Essentially two schools of thought can be identified.

One school, represented by Model A (Figure 2), considers that drainage shortens the duration of the peak flow; that is, base time is

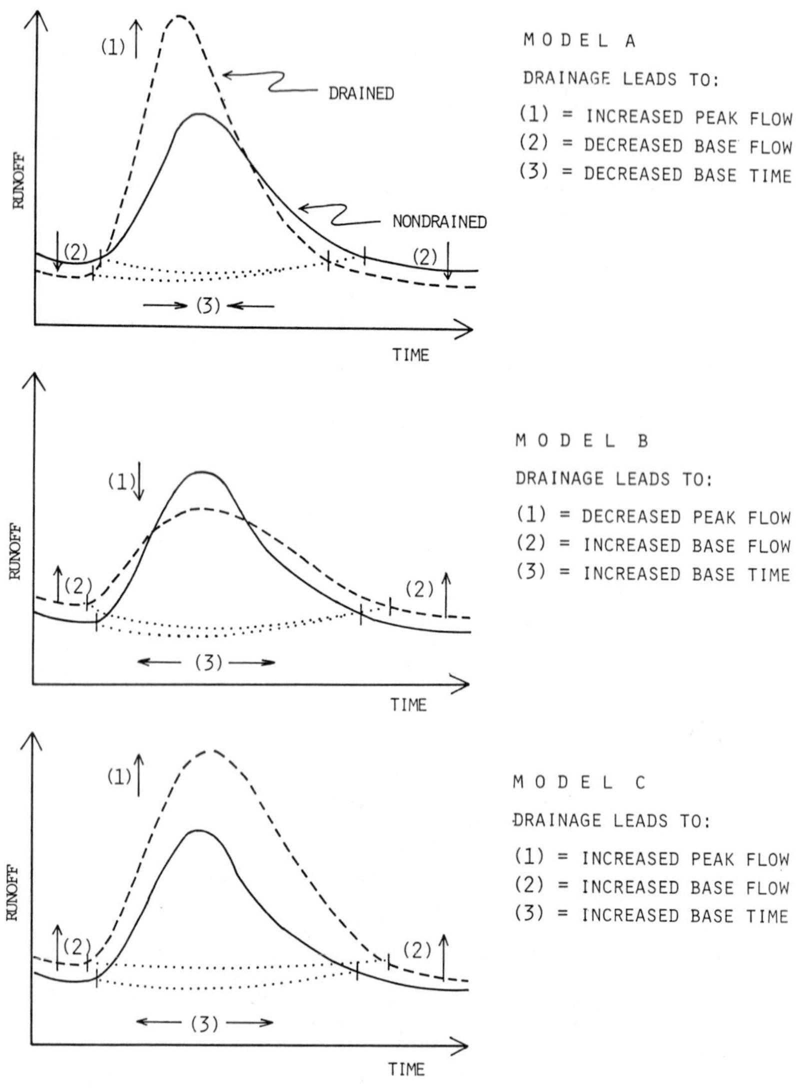

Figure 2. Hydrograph models (based on McDonald 1973).

reduced, peak flow levels are increased, and base flow is decreased. Surface runoff plus subsurface runoff (= direct runoff) brings water to the ditch network, which has an accelerating effect on the discharge from the catchment. A less stable, "flashy" hydrograph with accentuated peak flows is thus observed (e.g., Mustonen and Seuna 1971; Ahti 1980).

The second school, represented by Model B (Figure 2), considers that peak flow levels are reduced as a result of drainage, but the duration of the peak flow event is extended and base flow levels are increased. The effect of a more uniform flow over time that results is to "level out" the hydrograph. Exponents of this model include Multamäki (1962), Heikurainen (1976 and 1980), and, in Ireland, Burke (1972).

Similar models have been described by McDonald (1973), but a third model (Model C, Figure 2), in which both the peak flow and base flow levels are increased, may also exist. Such a model would clearly increase the total volume of runoff (discharge) after drainage. Seuna (1974 and 1980), comparing hydrographs from drained and nondrained peatland dominated watersheds, observed an increase both in the maximum flow levels during the highest flow events and in the minimum flow levels. There was, however, a tendency for these increases to decline each year after drainage.

THE INFLUENCE OF CATCHMENT CHARACTERISTICS ON RUNOFF

Other than slope, topography, and area of the site involved, the particular characteristics of the peatland catchment have been found to affect the response of runoff after drainage include site type, peat type, intensity of drainage, and type of ditch used. To some extent the influences of slope and topography are incorporated in the site type and peat type factors, since the growing vegetation and the associated underlying peat are related to habitat conditions, which are partly determined by topography and slope. Because of differences in the composition and structure of vegetation and the presence or absence of a stand, evapotranspiration and interception capacities vary with site type, and consequently so does runoff. Different peat types have different infiltration capacities and hydraulic conductivities (e.g., Päivänen 1973), and thus will react differently to drainage (McDonald 1973).

Water, once in the ditch network, is capable of being removed rapidly from the catchment. Therefore, the potential runoff flow is determined by the character of the ditch network itself--that is, the position, orientation, and condition of the ditches. The design of the ditch network depends on site type, slope, and topography of the drainage area, and the location of a suitable runoff outlet.

The discharge from a drained catchment tends to be inversely related to the ditch spacing used (e.g., Huikari 1963). By reducing ditch spacings, runoff distances to the nearest ditch are correspondingly reduced and the hydraulic gradient within the peat steepened. Thus, Ahti (1980) found monthly maximum peak flows from a drained open catchment were inversely related to ditch spacings. Increasing the density of ditches also increases the amount of precipitation that is collected directly by open ditches.

A further consideration of the drainage network is the type of ditch used. Päivänen (1976) has examined the effects of three different types of contour ditch (ordinary open ditches, covered plastic pipe drains, and narrow vertical-walled ditches) on the hydrology of an open bog. The results showed that the highest peak flows were recorded from catchments drained with the ordinary open ditches, which are used in practice almost exclusively.

THE INFLUENCE OF PRECIPITATION CHARACTERISTICS ON RUNOFF

The response of the peak flow to drainage is dependent upon the intensity and duration of rainfall. Most researchers make a distinction between the runoff response during "dry" and "wet" periods. Runoff during a "dry" period is largely made up of base flow runoff, while runoff during a "wet" period is dominated by peak flow runoff.

During periods of low rainfall, the utilization of the water storage capacity of the surface peat layer and the interception of the tree stand are maximized. Drainage leads to an increase in the interception capacity through stimulated tree growth. Changes in the water storage capacity are, however, disputable. For example, Heikurainen (1976) considers the water storage of the surface peat layer to have increased after drainage because of a lowering of the water-table level. The surface peat layer acts as a buffer; initially absorbing incoming precipitation, and thus reducing peak flow levels (Model B, Figure 2). Heikurainen (1976) attributes the increased base flow in Model B to the influence of the "runoff threshold" concept--that is, the distance to the water table at which direct runoff ceases. After rainfall the water table subsides and falls below the runoff threshold depth. Since the runoff threshold lies nearer the surface in nondrained peatlands, runoff tends to cease sooner from a nondrained area than from a drained area.

The utilization of the water storage capacity in the peat depends, however, on the rate of the incoming precipitation in relation to the infiltration rate of the peat. The infiltration capacity and hydraulic conductivity of the peat are probably reduced after drainage because of subsidence and enhanced decomposition (McDonald 1973). Seuna (1980) states that infiltration is, however, slightly increased after drainage, but implies that the storage capacity of the peat can only be used in light rainfall events. Thus, Seuna (1974 and 1980) observed that the lower peak flow levels did not increase after drainage partly because of the utilization of the storage capacity of the peat layer.

The increased soil storage and stand interception capacities due to drainage are considered by Heikurainen (1976) to be effective in absorbing normal average rainfall events. These capacities are lower on nondrained catchments, especially open peatlands, and are rapidly filled, resulting in a quicker and more intense runoff response compared with drained catchments. However, Heikurainen (1976) did observe higher peak flow levels from drained forested catchments after extreme rainfall. During such events the interception capacity of the stand and the infiltration capacity of the peat surface layer are exceeded, and the more rapid-moving direct (surface plus shallow subsurface) runoff occurs. The higher peak flow levels recorded are then due to the accelerating effect of the ditches which collect the runoff (Seuna 1974 and 1980). Thus, after drainage, the peak flow caused by the heavy rainfall events tends to be increased the most, in relative terms (Seuna 1974 and 1980).

Seuna (1974 and 1980) attributes increased minimum flows to the presence of the ditches, which make flow from the catchments possible at all times. The main ditches in the research basin reached the underlying pervious mineral soil, enabling input into the ditch from the ground water table. Clearcutting of peatland forest has been shown to increase catchment runoff (Päivänen 1974), and therefore harvesting carried out on the research catchments during the postdrainage period may also have led to the increased base flows. Nevertheless, Seuna (1974 and 1980) observed that the increase in base flow after drainage showed a significant tendency to decline over time. This decline was associated with the concurrent increases in stand interception and evapotranspiration.

THE SPRING PEAK HYDROGRAPH

The frequency of summer flooding is low, perhaps once in twenty to fifty years. High water-table levels in the peat during the summer have been shown to have a detrimental effect on root growth (Vompersky 1968). Under Finnish conditions, water tables artificially maintained near the soil surface during July and August have been shown to reduce the growth of Scots pine (Pelkonen 1975; Päivänen 1984). However, flooding for only a few hours is sufficient to induce anaerobic conditions in the soil and cause serious root damage (Coutts and Armstrong 1976).

Spring floods, although occurring perhaps once every two years, take place before the growing season and thus cause little damage directly to agricultural crops or tree stands compared with summer floods. Nevertheless, the size and frequency of spring flooding can result in serious damage to buildings, roads, and bridges, and may lead to the progressive deterioration of soils through leaching.

The characteristics of rainfall induced peak flows are determined by the balance that is achieved between precipitation, interception and evapotranspiration, infiltration, and water storage. The character of the spring runoff peak, however, is determined mainly by the water equivalent of snow, the water storage capacity of the soil, and temperatures during snowmelt.

Advocates of hydrograph Model A envisage drainage as having a similar effect on the spring peak flow as summer peak flows--that is, intensified peak flows. But it may be argued that because of the large quantities of water released during snowmelt, runoff from catchments would behave independently of the presence or absence of drainage. Drainage would then have a relatively smaller effect on the spring peak flow compared with summer peak flows. Thus, Seuna (1974 and 1980) found that, during the first nine years of drainage, the spring maximum flow from a sparsely forested catchment increased, on average, by 31%, while the corresponding value for the summer maximum flow was 131%. Nevertheless, the occurrence of flooding is related to absolute peak flow values.

Advocates of Model B, on the other hand, argue that drainage decreases spring peak flows. Heikurainen (1976) attributes the decrease to a delay in snowmelt, caused by the presence of the tree cover, and to a water storing zone that lies between the lower limit of soil frost and the water table. This zone is considered to be completely lacking in nondrained peatlands.

DISCUSSION AND CONCLUSIONS

The apparent discrepancy between the various studies dealing with the hydrological impact of peatland drainage for forestry purposes has arisen largely because of (1) differences in catchment characteristics, primarily whether it is an open or forested catchment, (2) differences in rainfall/snowmelt characteristics, and (3) time since ditching.

The three hydrological models presented in Figure 2 are not necessarily mutually exclusive, but rather long-term phases of post-drainage development. Model C probably represents the situation immediately after drainage, especially in the case of open peatlands, when evapotranspiration is substantially reduced because of the lowering of the water table and loss of the hydrophytic vegetation cover. Peak and base flow levels are thus increased. However, as the stand grows, interception and evapotranspiration begin to increase and Models A and B, or combinations, are more probable.

Random variation about this long-term declining trend in runoff is caused by the constantly varying characteristics of precipitation. For example, during heavy rainfall events, there is a tendency for greater peak flow levels from drained catchments because of the accelerating effect of the ditches (i.e., Model A or C). During light or normal rainfall events, however, the increased interception and water storage capacity of the soil are effective and peak flow levels from drained catchments reduced (i.e., Model B).

Drainage thus increases the potential for greater runoff peak flow levels from catchments. Whether this potential is realized, however, depends on the density of ditches, the characteristics of the rainfall/snowmelt, and, in the longer term, the growth of the stand. Although the area of drained peatland in Finland for forestry purposes is extensive, the ditching densities used are low when compared with those used in agriculture or fuel peat production. Furthermore, most of the peatland area that has been drained had a natural forest cover at the time of ditching, the peak flow stabilizing effect of which increases as stand growth responds to drainage. Finally, the initial potential for increased peak flows will become progressively of less importance, for the ditching of new peatland areas has now virtually ceased.

REFERENCES

Ahti, E. 1980. Ditch spacing experiments in estimating the effects of peatland drainage on summer runoff. Proc. Helsinki Symp., pp. 49-53. IAHS-AISH Publ. 130.

Baden, W., and R. Eggelsmann. 1968. The hydrologic budget of the highbogs in the Atlantic region. Proc. Third International Peat Congress, Quebec, pp. 206-211.

Burke, W. 1972. Aspects of the hydrology of blanket peat in Ireland. Int. Symp. Hydrology of Marsh-Ridden Areas, Minsk, Byelorussian SSR. 16 pp.

Coutts, M. P., and W. Armstrong. 1976. Role of oxygen transport in the tolerance of trees to waterlogging. In: M. R. G. Cannel and F. T. Last, eds., Tree physiology and yield improvement. Academic Press, New York.

Green, F. H. W. 1973. Hydrology in relation to peat sites. In: Peatland forestry, pp. 103-106. NERC, Edinburgh.

Heikurainen, L. 1976. Comparison between runoff conditions on a virgin peatland and a forest drainage area. Proc. Fifth International Peat Congress, Poznan, 1:76-86.

__________. 1980. Effect of forest drainage on high discharge. Proc. Helsinki Symp., pp. 89-96. IAHS-AISH Publ. 130.

Huikari, O. 1963. Effect of distance between drains on the water economy and surface runoff of Sphagnum bogs. Proc. Second International Peat Congress, Leningrad, 2:739-742.

Hyvärinen, V., and B. Vehviläinen. 1978. The influence of forest draining on discharges in Finland. Proc. Nordic Hydrological Conference, Hanasaari, Helsinki, 3:1-10.

McDonald, A. 1973. Some views on the effect of peat drainage. Scottish Forestry 27(4):315-327.

Multamäki, S. E. 1962. Die Wirkung von Waldentwässerung auf die Ablaufverhältnisse von Torfboden. Commun. Inst. For. Fenn. 55(23): 1-16.

Mustonen, S. E., and P. Seuna. 1971. Metsäojituksen vaikutuksesta suon hydrologiaan. (Influence of forest draining on the hydrology of peatlands.) National Board of Waters, Finland, Water Res. Inst. Publ. 2, pp. 1-63.

Päivänen, J. 1973. Hydraulic conductivity and water retention in peat soils. Acta For. Fenn. 129:1-70.

__________. 1974. Hydrological effects of clear cutting in peatland forests. Proc. Int. Symp. Forest Drainage, Jyväskylä-Oulu, Finland, pp. 219-228.

__________. 1976. Effect of different types of contour ditches on the hydrology of an open bog. Proc. Fifth International Peat Congress, Poznan, 1:93-106.

__________. 1984. The effect of runoff regulation on tree growth on a forest drainage area. Proc. Seventh International Peat Congress, Dublin. In press.

Pelkonen, E. 1975. Vuoden eri aidoina korkealla olevan pohjaveden vaikutus männyn kasvuun. (Effects on Scots pine growth of ground water adjusted to the ground surface for periods of varying length during different seasons of the year.) Suo 26(2):25-32.

Seuna, P. 1974. Influence of forest draining on the hydrology of an open bog in Finland. Proc. Int. Symp. Forest Drainage, Jyväskylä-Oulu, Finland, pp. 385-393.

__________. 1980. Long-term influence of forestry drainage on the hydrology of an open bog in Finland. Proc. Helsinki Symp., pp. 141-149. IAHS-AISH Publ. 130.

Vompersky, S. E. 1968. Biologicheskie osnovy effektivnosti lesoosusheniia. (Biological foundations of forest drainage efficiency.) Akademiia Nauk SSSR. "Nauka," Moscow. 311 pp.

SITE CLASSIFICATION AND PREDICTION IN YOUNG CARIBBEAN PINE PLANTATIONS IN GRASSLANDS OF VENEZUELA

L. W. Vincent

The government of Venezuela, through the combined efforts of the state-owned Guayana Corporation (CVG) and the National Reforestation Company (CONARE), is in the process of establishing large-scale industrial plantations on natural grasslands in eastern Venezuela. The main objectives are to produce long-fiber pulp and timber for particleboard and lumber. Current annual planting objectives for the Uverito (CVG) and other projects (CONARE) are approximately 20,000 hectares, mostly with Caribbean pine.

The present study was designed to derive a site classification system for young Caribbean pine (*Pinus caribaea* Mor. var. *hondurensis* Barr. et Golf.) in order to provide (1) a basis for decision on which areas to plant, (2) a stratification for yield estimates, (3) a basis for decisions regarding stand treatment, and (4) a set of references for research purposes.

THE STUDY AREA

The study was carried out in Uverito in eastern Venezuela, just north of the Orinoco River and Puerto Ordaz (8°30' to 8°48'N, 62°30' to 63°8'W).

The grasslands being planted are situated at around 20 m above sea level on relatively flat terrain. The annual precipitation is around 900 to 1,000 mm. The main rainy season occurs from June to August, and a minor one sporadically occurs in November and December. Estimates of mean annual temperature range from 25 to 26.5°C. Relative humidity is quite high during most of the year, with nocturnal highs of up to 100% even during the driest months.

The grasslands soils are for the most part sandy and of low fertility. Land use has been mainly extensive cattle grazing of marginal economic value. For this reason there was little problem in setting aside these large areas for forestry development.

Despite the first impression of a seemingly uniform site, there is considerable variation in soil properties, especially regarding depth of the sandy A horizon. The variation is due to the complex geological history of the area (Mercier 1976). The parent material was deposited during the late Pliocene to early Pleistocene in a continental near shoreline environment. Subsequently there took place a period of erosion resulting in the cutting of minor valleys, followed by aeolic reorganization under a drier climate, resulting in varying depths of sand in the form of dunes and filled-in valleys, and exposed hardpanned B horizons.

The grassland vegetation is made up mostly of a coarse *Trachypogon* grass and scattered shrub-size *Curatella americana* and other species. The grassland mesas are dissected by the "morichales," which are streams with surrounding swampy gallery forest dominated by the moriche palm (*Mauritia minor*).

MATERIALS AND METHODS

The original site classification study was carried out in 1977-78 (Vincent 1978) and has been updated with data taken in 1981 (Chaves 1982) and 1982.

Field sampling was done under two major categories: (1) a 102-plot block-stratified random sample was taken in a plantation of uniform age (six years in 1977 and 10.3 years in 1981) over a 3,000 hectare area in order to derive the classification schemes and obtain site factor and stand data for correlation and regression analyses; and (2) permanent sample plot and stem analysis data were used to derive a dominant height-age regression model for the site curves.

Two dominant height site index variables were used. Most site classification work in Venezuela has been based on a fixed area plot estimate of dominant height (DHf), in which the height of the tallest tree in each 100-square-meter subplot is measured. DHf is the plot average of such measurements. However, in poorly stocked stands many such subplots were found to be vacant or with so few trees that the ones measured were not dominant or otherwise were unfit for measurement (usually "foxtails"). Hence in the original study, variable-size plots, in which dominant height (DHv) at age six was taken as the average height of the ten well-spaced dominants nearest plot center, were used to overcome this difficulty. The radii of the plots varied inversely with stand density. Consequently, in a poorly stocked stand a greater plot radius was required to select the ten dominants than in more adequately stocked stands.

Despite its problems with low density stands, DHf was used in the recent update of the site classification scheme to provide a basis for use of the more common site index expression. DHf at age 10.3 was measured on the same sample points established in 1977. Circular plots, subdivided into eight 120-square-meter subplots, were used (Chaves 1982).

Additional stand variables included basal area measured with a five-factor metric prism based on an average of 17 points per plot, and average girth and distance of dominants from plot center.

A field soil description and samples were taken for each plot at plot center using a soil auger. Slope was measured in two directions using a hand level, and topographic position was registered using a scale of five classes, later simplified to three.

For the dynamic aspect of the study, permanent plot DHf data for the period 1974-82 and stem analysis data taken in 1977-78 were used to derive the DH-age model. Validity of the stem analysis data was established in the original study. Problems with false rings were overcome through ring dating techniques. The stem analysis data were used to supplement permanent plot data in both the original and update studies.

DATA PROCESSING AND ANALYSES

The major steps followed in processing the data are summarized as follows: (1) testing of various classification schemes with different numbers of categories and intervals of DHv through trial and error with distribution of plots as criterion; (2) correlation and regression analyses using DHv as the dependent variable and various sets of site factor variables as independent variables; (3) multiple discriminant analyses and classification to test the discriminating power of various sets of site factor variables (test batteries) among site classes defined in step 1, and to derive a predicting function; (4) regression of DH on age to derive a model for the DHv and DHf site classification curves; and (5) map-testing of the classification scheme using an intensively systematically sampled 54-hectare block.

Derivation of the classification schemes was based on the sample means, standard deviations, and ranges of DHv and DHf (step 1). The DHv age six scheme was based on 97 of the original 102 plots after rejection of five plots with extremely low DHv values which fell outside the general distribution. The DHf age 10.3 scheme was derived after rejection of plots having less than four valid subplot dominant tree measurements. A valid subplot was taken to have at least the equivalent of 50 trees per hectare. Only 56 of the original 102 plots met these requirements.

A stagewise procedure was used in steps 2 and 3, with three sets of independent site factor variables. The sets were based on the degree of difficulty and cost of obtaining data. DHv at age six was the dependent variable in regression analyses, and DHv site class (also base age six)

Table 1. Description of soil-site independent variables used in stage 1 statistical analyses.

Variable Number	Description	Units
FX_1	Depth to textural change from sand to sandy loam or heavier texture	cm
FX_2	Depth to textural change from sand to sandy clay loam or heavier texture	cm
FX_3	Depth of "organic layer" based on color	cm
FX_4	Average of absolute values of slope	Degrees
FX_5	Algebraic sum of slopes	Degrees
FX_6	Topographic position (3 positions: flat, dome, bottom) expressed as 2 dummy variables	Categories
FX_7	Depth to mottling	cm
FX_8	Depth to red soil	cm

Table 2. Description of soil-site independent variables used in stage 2 statistical analyses.

Variable Number	Description	Units
FX_1-FX_8	Variables used in stage 1	
X_1	Depth to where sand content is less than 80%	cm
X_2	Depth to where sand content is less than 85%	cm
X_3	Depth to where sand content is less than 90%	cm
X_4	% sand at 50 cm	%
X_5	% sand at 75 cm	%
X_6	% sand at 90 cm	%
X_7	% sand at 100 cm	%
X_8	% sand at 125 cm	%
X_9	% sand at 150 cm	%
X_{10}	% sand at 200 cm	%
X_{11}	% sand in stratum where sand # 80% (X_1)	%
X_{12}	Weighted average of % sand in 50 cm below X_1	%
X_{13}	Weighted average of % sand down to 50 cm	%
X_{14}	Weighted average of % sand down to 100 cm	%
X_{15}	Weighted average of % sand down to 150 cm	%
X_{16}	Weighted average of % sand down to 200 cm	%

was the criterion variable in discriminant analyses. Stage 1 variables were those resulting exclusively from field descriptions (Table 1). Stage 2 variables included those in stage 1 plus variables computed from laboratory texture analyses (Table 2). Stage 3 variables included those of stages 1 and 2 plus others as indicated in Table 3.

In each stage, variables were first screened according to simple correlations with DHv and among themselves to avoid colinearity. Next, backward, forward, and maximum R^2 stepwise multiple regression procedures were used to select variables and derive predicting models. All

Table 3. Description of soil-site independent variables used in stage 3 statistical analyses.

Variable Number	Description	Units
FX_1-FX_8	Variables used in stage 1	--
X_1-X_{16}	Variables used in stage 2	--
X_{17}-X_{21}	Phosphorus at specified depths[a]	ppm
X_{22}-X_{26}	Weighted average of phosphorus down to specified depths	ppm
X_{27}-X_{31}	Potassium at specified depths	mill.eq./100 g
X_{32}-X_{36}	Weighted average of potassium down to specified depths	mill.eq./100 g
X_{37}-X_{41}	Organic matter at specified depths	%
X_{42}-X_{46}	Weighted average of organic matter down to specified depths	%
X_{47}	Moisture-holding capacity 1/3 atm at 90 cm depth	%
X_{48}	Moisture-holding capacity 15 atm at 90 cm depth	%

[a]Specified depths used in analyses are 10, 15, 25, 35, and 45 cm.

statistical analyses in the original study were carried out using SAS (Statistical Analysis System; SAS Institute 1979) and the Cooley and Lohnes (1971) MANDVA, DISCRM, and CLASIF subroutines on an IBM 370 computer. The number of observations used depended upon limitations imposed by missing data.

Separate polynomial individual site factor model regression analyses were used to test fitting of the best independent variables. This procedure was used to determine the validity of multiple transformation expressions of variables in the combined regression models.

Seven models were fitted in step 4 using the updated DH-age permanent plot and stem analysis data with 90 observations on an Apple II+ microcomputer. The best model was selected on the basis of R^2 and simplicity.

In step 5, DHf at age six was measured on each of the 126 plots in the intensively sampled 54-hectare block. This sampling was done at the beginning of the original study and provided the basis for the decision to use the new variable plot DH, owing to problems in understocked stands. A regression model was derived to convert the DHv site curves to permit classification of the plots according to the DHv scheme. The plot locations with corresponding DHv site class designation were plotted to test spatial distribution.

RESULTS

A four-class DHv site classification scheme was selected in the original study, in which the intervals are defined in terms of the mean and standard deviation of DHv for age six (Table 4). A new three-class scheme based on DHf was derived from the 1981 data (10.3-year base age) using the 56 valid values. Class intervals were determined on the basis of the sample mean and range. Interval separators were:

Classes I/II $\overline{X} + 1/3(\max - \overline{X})$

Classes II/III $\overline{X} - 1/3(\overline{X} - \min)$

Model 1 was selected from seven height-age models tested based on R^2 for the updated data.

$$\text{Ln DH} = 3.2591 - 6.7328(1/A) \qquad (1)$$

where Ln = natural logarithm
DH = dominant height (m)
A = age (yrs)

with $R^2 = 0.95$
n = 98 observations

The resulting site classification curves are given in Figures 1 and 2 for DHv and DHf. Figure 3 demonstrates the spatial distribution of the DHv site classes.

Multiple regression analyses of DHv on site factor variables gave R^2 values of .52 for a 12-variable model in stage 1, .60 for a 10-variable model in stage 2, and .68 for a 13-variable model in stage 3 (Tables 5, 6, and 7). The largest R^2 value was .87 for a 44-variable model in stage 3. Stage 1 single factor models gave R^2 values of .21 for FX6 (topographic position, two dummy variables) and .26 for a fourth degree polynomial of FX1 (depth to textural change from sand to sandy loam or heavier). Table 8 gives simple correlation coefficients for stage 1 variables and some of their transformations, as an example to show how the screening of variables was done at each stage.

The results of discriminant analyses for the three stages are summarized in Table 9. Only the first discriminant function is significant in all three stages. The null hypothesis for equality of centroids was rejected (.01 significance level). Percentages of misclassified plots varied from 11 for stage 3 to 42 for stage 1. The site variables included in the selected models for each stage are given in Table 10. Model type refers to how the variables were selected for inclusion in the test batteries. "Regression" means that variables included were those found best in regression analyses.

The map of centroids in the discriminant subspace defined by the first and second discriminant functions (Figure 4) indicates the greater discriminating power of the first function, except between site classes II and III. Discrimination between the latter is better along the second function.

Table 4. Selected site classification scheme class interval specifications and sample statistics at age six.

Site Class	Interval: Std. Dev.	Interval: DH_v (m)	Frequency of Plots (no.)	Frequency of Plots (%)	Means[a] BA (m^2/ha)	Means[a] D (cm)	Means[a] dbh (cm)	CV BA (%)	CV D (%)	CV dbh (%)
I	$DH_v > (\overline{X} + s)$	> 10.8	13	12.8	13.68	15	18.3	26	22	5
II	$\overline{X} < DH_v \leq (\overline{X} + s)$	10.15-10.8	40	39.2	10.90	17	17.9	42	23	6
III	$(\overline{X} - s) < DH_v \leq \overline{X}$	9.50-10.15	29	28.4	7.44	18	17.9	59	25	6
IV	$DH_v \leq (\overline{X} - s)$	≤ 9.5	20	19.6	2.55	30	16.0	86	66	9
Totals			102	100.0						
Analysis of variance F ratio[b]					27.26	10.49	17.04			

[a]BA = basal area; D = average distance of selected dominant trees from plot center; dbh = diameter breast height.

[b]All F ratios are significant at a = .0001.

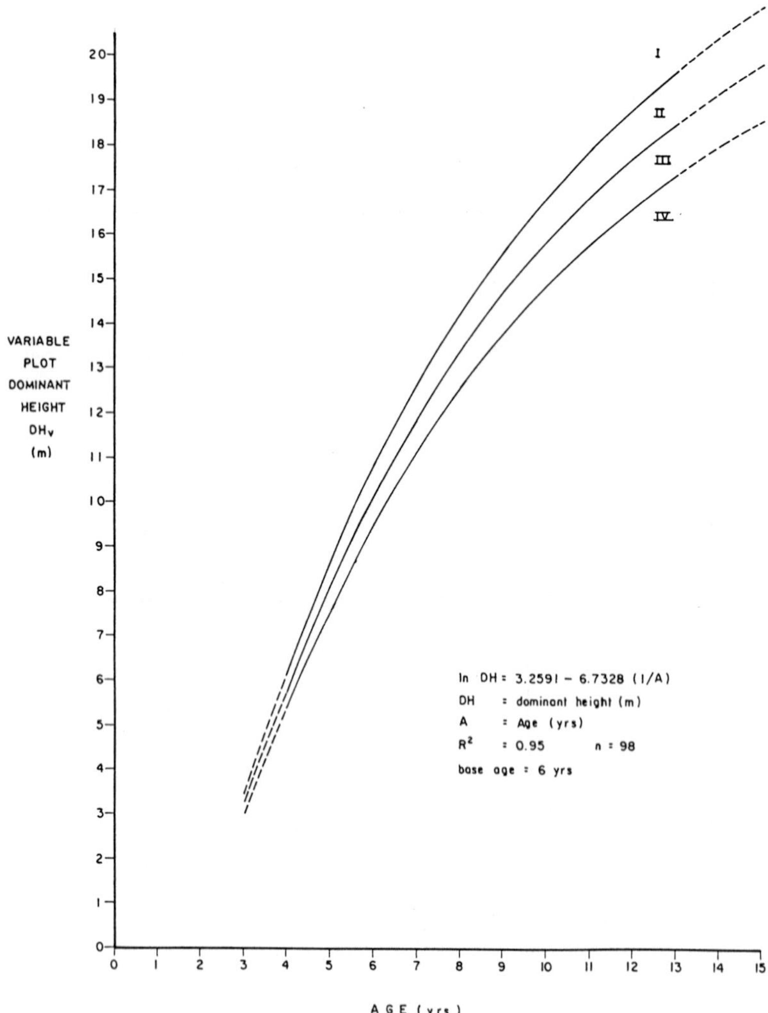

Figure 1. Variable plot dominant height (DHv) site classification curves based on the original base age six scheme and the updated dominant height-age regression model.

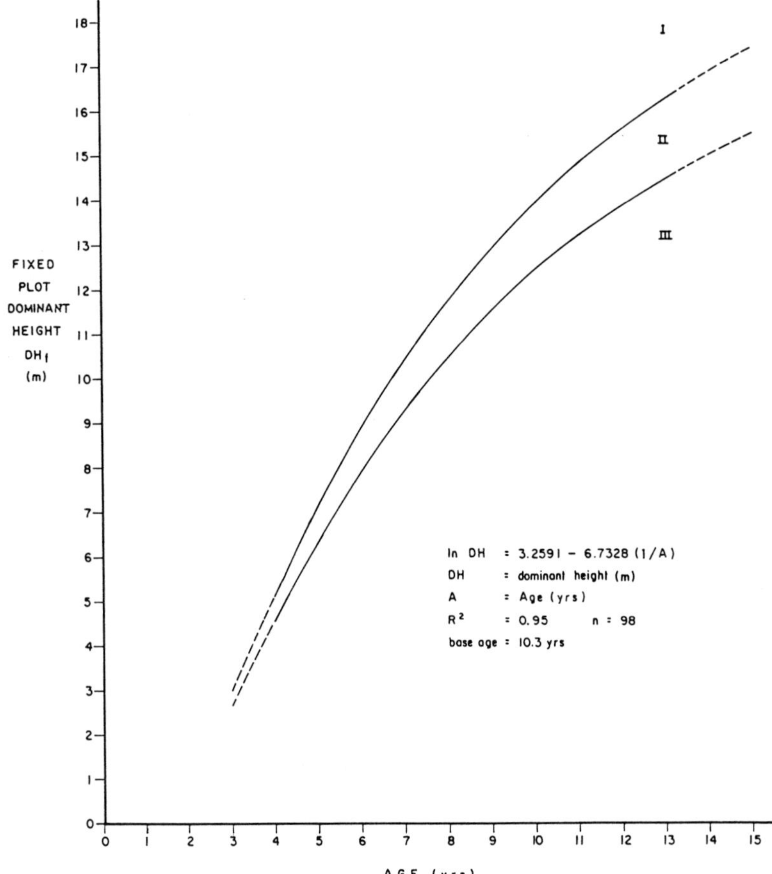

Figure 2. Fixed plot dominant height (DHf) site classification curves based on the updated study base age 10.3 scheme and the updated dominant height-age regression model.

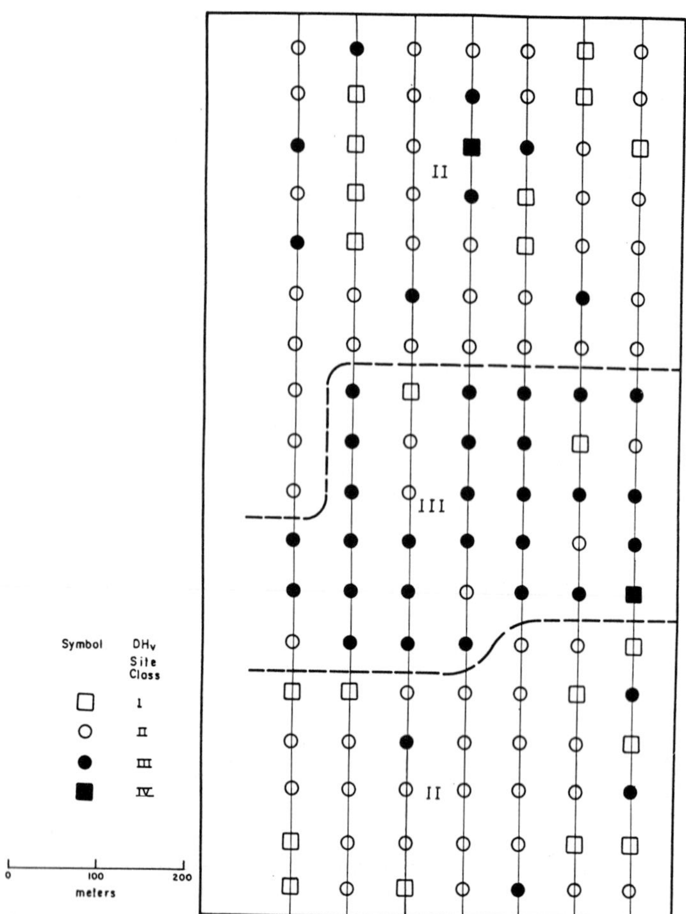

Figure 3. Spatial distribution of DHv site classes on a 54-hectare intensively systematically sampled block at age six.

Table 5. Stage 1 selected regression model descriptions and statistics.

Model Number	Variables[a] (no.)	Factors[b] (no.)	Largest Number of Transformations of a Variable (no.)	Regression Statistics R^2	F^c	Degrees freedom error (no.)	S_e (m)	Smallest partial contribution sig.
15	6	5	1	.31	6.03	82	.72	.5712
16	9	5	2	.39	5.58	79	.69	.7505
17	30	5	7	.69	4.85	66	.62	.5999
18	25	5	7	.68	5.97	71	.61	.2242
19	13	5	4	.51	6.71	83	.69	.1617
20	12	5	5	.52	7.52	84	.68	.2051

[a] Variables include all transformations computed from a factor (raw material).

[b] Factors raw material (untransformed).

[c] All F ratios are significant at a = .0001.

Table 6. Stage 2 selected regression model descriptions and statistics.

Model Number	Variables (no.)	Factors (no.)	Largest Number of Transformations of a Factor (no.)	Regression Statistics R^2	F^a	D.F. error (no.)	S_e (m)	Smallest[b] partial contribution sig.
26	37	8	8	.76	4.26	51	.57	.5597
27	18	8	5	.64	6.77	70	.59	.5679
28	13	8	2	.60	8.81	75	.59	.4994
29	10	7	2	.60	11.58	78	.59	.4390
30	5	4	1	.55	20.01	83	.60	.0110

[a] All F ratios are significant at a = .0001.

[b] Least significant contribution other than $FX_{6(2)}$--second topographic position dummy variable.

Table 7. Stage 3 selected regression model descriptions and statistics.

Model Number	Variables (no.)	Factors (no.)	Largest Number of Transformations of a Factor (no.)	Regression Statistics				
				R^2	F^a	D.F. error (no.)	S_e (m)	Smallest[b] partial contribution sig.
31	44	12	8	.87	6.47	44	.45	.4281
32	25	10	4	.75	7.33	62	.52	.3207
33	13	8	3	.68	11.87	74	.54	.1844
34	11	8	3	.66	13.69	76	.54	.0603
35	15	9	3	.69	10.86	73	.53	.4741
36	13	8	3	.68	12.43	75	.53	.3343
37	6	5	1	.54	16.42	82	.61	.0518
38	7	6	1	.56	14.56	81	.60	.1525

[a] All F ratios are significant at a = .0001.

[b] Least significant contribution other than $FX_{6(2)}$--second topographic position dummy variable.

Table 8. Simple correlation coefficients for stage 1 independent variables with DH_v and six largest correlations with other independent variables.

Variable[a]	Correlation with DH_v[b]	Strongest Six Correlations with Other Site Factor Variables											
		1st		2nd		3rd		4th		5th		6th	
		Var.	r	Var.	r	Var.	r	Var.	r	Var.	r	Var.	r
FX_8[c]	-.50	FX_1	.85	FX_1^2	.81	FX_2	.69	FX_7	.52***	FX_3	.30**	FX_5^2	.28**
FX_1	-.47	FX_8	.85	FX_2	.77	FX_2^2	.74	FX_7	.59	FX_5^2	.34***	FX_4	.30**
FX_2	-.38	FX_7	.90	FX_1	.77	FX_1^2	.69	FX_8	.69	FX_4	.40	$FX_{6(1)}$	.38
FX_7[c]	-.37**	FX_2	.90	FX_2^2	.88	FX_1	.59	FX_1^2	.55	FX_8	.52***	FX_4	.40**
FX_3	.32***	FX_1	-.20*	$FX_{6(1)}$	ns	FX_8	ns	FX_2^2	ns	FX_1^2	ns	FX_2	ns
FX_4	-.21*	FX_5^2	.51	FX_2^2	.42	FX_2	.40	FX_7	.40**	FX_1^2	.30**	FX_8	.30**
FX_5	ns	$FX_{6(2)}$	.23*	FX_5^2	ns	$FX_{6(1)}$	ns	FX_4	ns	FX_8	ns	FX_7	ns
FX_1^2	-.44	FX_8	.81	FX_2^2	.73	FX_2	.69	FX_7	.55	FX_5^2	.40	$FX_{6(2)}$	.30**
FX_5^2	-.39	FX_4	.51	$FX_4^{.5}$	.41	FX_1^2	.40	FX_2^2	.36***	FX_1	.34***	FX_2	.32***
FX_2^2	-.45	FX_7	.88	FX_1	.74	FX_1^2	.73	FX_8	.68	FX_4	.42	$FX_{6(1)}$	.41
FX_2^3	-.42	--	--	--	--	--	--	--	--	--	--	--	--
$FX_3^{.5}$	.35***	FX_8	ns	FX_1	ns	$FX_{6(1)}$	ns	FX_2^2	ns	FX_1^2	ns	FX_2	ns
$FX_4^{.5}$	ns	FX_5^2	.41	FX_7	.38**	FX_2^2	.35***	FX_2	.34***	$FX_{6(1)}$	.27**	FX_8	.26*
$FX_4^{2.5}$	-.27**	--	--	--	--	--	--	--	--	--	--	--	--

[a] First untransformed variables are listed followed by best transformations of each variable.

[b] All correlation coefficients are significant at α = .0001 unless otherwise indicated: ns α > .05: * α < .05; ** α < .01; and *** α < .001.

[c] Not included in multiple regression analyses due to missing values.

Table 9. Multivariate statistics for alternative discriminant models tested.

	Model Type										
	Regression				No transformation			Best transformation			Intuition
			Stage 3								
Multivariate Statistic[a]	Stage 1 (20)	Stage 2 (28)	(36)	(33)	Stage 1 (39)	Stage 2 (40)	Stage 3 (41)	Stage 1 (42)	Stage 2 (43)	Stage 3 (44)	Stage 3 (45)
Number of plots[b]	97	96	98	98	97	95	91	97	95	91	93
Number of variables	11	11	11	12	6	8	12	6	8	12	12
Wilk's Λ[c]	.3803	.3695	.3852	.3622	.5393+	.4705	.3888	.5563+	.4953	.3781	.4661
F for H: $u_k = u$[d]	2.89	2.95	2.88	2.80	3.38+	3.02	2.36	3.19+	2.79	2.44	1.89+
F for H: $\Delta_k = \Delta$[e]	2.36	2.61	2.42	3.84	--	1.84	1.73	--	2.13	1.64	--
+χ^2 for H: $\Delta_k = \Delta$[f]	536	659	859	995	123	214	465	156	246	425	483
χ^2 tests for) 0[g]	--	--	85	92	--	66	77	--	--	--	--
successive) 1	--	--	18 ns	18 ns	--	12 ns	17 ns	--	--	--	--
roots removed) 2	--	--	4	7	--	3	3	--	--	--	--
% trace for) 0	--	--	84.2	85.7	--	86.1	83.1	--	--	--	--
successive) 1	--	--	12.1	8.8	--	10.8	13.9	--	--	--	--
roots removed) 2	--	--	3.6	5.5	--	3.1	3.0	--	--	--	--
canonical r) 0	--	--	.727	.752	--	.680	.722	--	--	--	--
for successive) 1	--	--	.373	.343	--	.312	.393	--	--	--	--
roots removed) 2	--	--	.215	.278	--	.175	.194	--	--	--	--
+% misclass. plots	26	17	22	24	42	32	11	43	33	11	15

[a] Statistics with + are from SAS analyses; others are from Cooley and Lohnes (1971).

[b] In all cases the smallest group is site class I with 13 plots and no missing values.

[c] Cooley and Lohnes (1971).

[d] Test of null hypothesis of equality of means. All F ratios are significant at $\alpha = .01$.

[e] Test of null hypothesis of equality of dispersions. All F ratios are significant at $\alpha = .01$.

[f] All χ^2 values are significant at $\alpha = .0001$.

[g] All "0 roots removed" χ^2 values are significant at $\alpha = .001$.

Table 10. Soil-site variables included in the discriminant models for stages 1 to 3.

Stage 1 (39)	Stage 2 (40)	Stage 3 (41)
FX_6	FX_6	FX_6
FX_1	FX_1	FX_1
FX_2	FX_2	FX_2
FX_3	FX_3	FX_3
FX_4	FX_4	FX_4
FX_5	FX_5	FX_5
--	X_3	X_3
--	X_6	X_6
--	--	X_{47}
--	--	X_{20}
--	--	X_{37}
--	--	X_{30}

DISCUSSION

A problem was encountered in the new DHv site curves regarding the classification of a few sample plots measured in the same population in June 1981. These plots were classified into class II at age six and class IV at age ten according to the updated curves. Addditional study is needed to update the DHv site class intervals, since the assumption of proportionality of the curves through the age six class intervals may not hold for more advanced ages. There is doubt whether a four-class scheme is justified due to the narrowness of the intervals and the results of the spatial distribution test (Figure 3). A relatively simple alternative would be a two-class scheme based on combination of classes I and II and classes III and IV, leaving only the present II/III separator. Furthermore, additional research is needed to test the new DHf classification scheme and to determine the most appropriate scheme for management and research purposes.

The stagewise procedure appears to be unique to this study, since no reference of similar work was found. The advantages of formulating alternatives of costs and precision seem to justify the data processing and statistical complexity of the procedure by making it possible for plantation managers and researchers to select the alternative most suitable to a given situation.

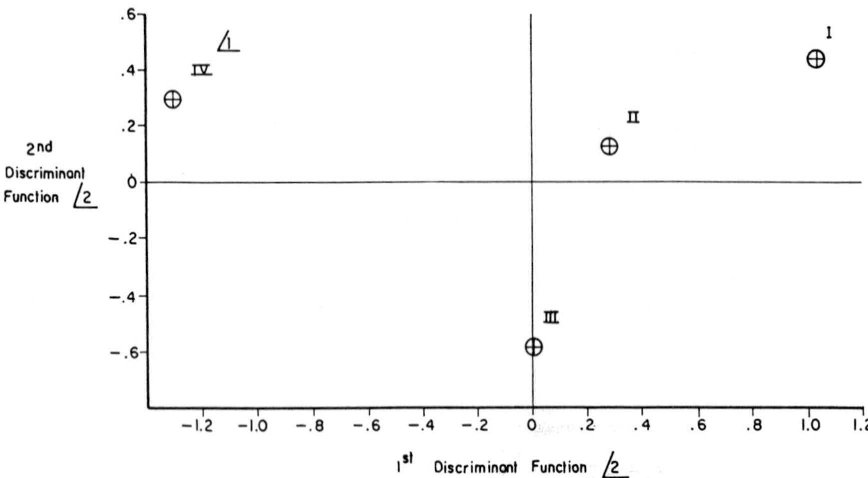

Figure 4. Variable plot base age six dominant height site class centroids on a two-dimensional discriminant function subspace.

[1] DHv site class.

[2] Standardized discriminant scores (unit variance, zero mean).

The results show that discriminant analysis constitutes a potentially useful predictive tool and that site classes can be differentiated on the basis of site factors. This tends to support the original four-class DHv scheme. As in the case of multiple regression, there was a gain in discriminatory power, hence predictive power, as additional variables were included from stages 1 to 3. The main conclusion from interpretation of the factor structure of the first discriminant function is that moisture-holding capacity related variables are predominant in determining site quality.

The relatively high R^2 values obtained in regression and the results of the discriminant analyses support the site classification system by showing that the DHv classification criterion used is related to site factors which should determine productivity, and that the defined site classes can be differentiated on the basis of site factors.

Obviously, additional study is needed to derive a more definite site classification system, since research involves two different site indices (DHv and DHf). Study is also needed of the problems regarding understocked stands and recently modified planting techniques that include intensive site preparation in contrast with areas that receive no site preparation prior to planting.

ACKNOWLEDGMENTS

The present study originated in a dissertation presented to the University of Tennessee (Vincent 1978). The author expresses his gratitude to the Consejo de Desarrollo Cientifico y Humanistico (CDCH) of the Universidad de Los Andes, Mérida, and the Corporacion Venezolana de Guayana, for financial support, to Dr. Frank Woods, as major professor, and to the University of Tennessee.

REFERENCES

Chaves, E. 1982. Relacion productividad-densidad de vuelo en plantaciones de _Pinus caribaea_ var. _hondurensis_ Barr. et Golf. en las sabanas al sur del estado Monagas, Venezuela. M.S. thesis, Universidad de Los Andes, Mérida, Venezuela.

Cooley, W. W., and P. R. Lohnes. 1971. Multivariate data analysis. John Wiley and Sons, New York.

Mercier, V. 1976. Estudio geomorfologico del area de Chaguaramas, Edo Monagas. Fondo Nacional de Investigaciones Agropecuarias, Ministerio de Agricultura y Cria, Maracay, Venezuela. Boletin Tecnico 3.

SAS Institute. 1979. SAS user's guide, 1979 edition. SAS Institute Inc., Raleigh, North Carolina.

Silva, R. 1971. Metodologia para investigacion en parcelas permanentes de clareo y rendimiento en plantaciones forestales. Boletin Instituto Forestal Latinoamericano de Investigacion y Capacitacion (IFLAIC)(Mérida) 38:59-89.

Vincent, L. W. 1978. Site classification for young Caribbean pine (_Pinus caribaea_ var. _hondurensis_) in grasslands, Venezuela. Ph.D. thesis, University of Tennessee (Dissertation Abstracts International Sec. B 39(11):5163, University Microfilms Order No. 7911723).

Voorhoeve, A. G., and J. P. Schulz. 1968. La necesidad de parcelas permanentes de clareo y rendimiento en plantaciones forestales. Boletin Instituto Forestal Latinoamericano de Investigacion y Capacitacion (IFLAIC) 27-28:3-17.

7

LAND CLASSIFICATION FOR INDUSTRIAL AFFORESTATION IN THE STATE OF SÃO PAULO, BRAZIL

M. A. M. Victor, F. J. N. Kronka, J. L. Timoni, and G. Yamazoe

INTRODUCTION

When a government intends to allocate resources to promote industrial afforestation, it is vital to classify the land in order to discipline this activity. Only through this instrument can one be sure that the subsidies destined to the sector will be used most effectively.

In the state of São Paulo a vigorous policy of fiscal incentives has been in effect since 1966 (Law 5.106 of 2/9/66), and two studies have been published (Secretaria da Agricultura 1970, 1975). Since these studies involve silvicultural and environmental considerations, besides economic and social ones, they are characterized by extreme dynamism and therefore demand periodic revision. A working team is currently undertaking a revision of the last study.

Ecological Concept of the State and the Remaining Vegetal Cover

The state of São Paulo, in southeast Brazil, is situated 44° to 53°W and 20° to 25°S, at altitudes varying from 0 to 2,242 m, with climates of the types Af, Aw, Cfa, Cfb, and Cwb (classification by Koeppen), with a predominance of precipitation in winter (typical in the south of Brazil) and alternating dry and wet seasons, including rainy summers. Hydric deficits occur only moderately in winter, in the central and eastern uplands and in the west (Ventura 1964). With an area of almost 25 million hectares (2.9% of the area of the country), the state has residual native forest equal to 8.3% of the territory (Victor 1975). In the mid-nineteenth century this cover equaled 81.8% of the total area, but the classical use of the land for forest, agriculture, and pasture reduced it to the present level. The maps in Figure 1 illustrate this phenomenon. Commercial afforestation, already more than seventy years old, has contributed, chiefly in recent years, to a substantial development equaling 2.58% of the area of the state. These data are shown in Table 1.

Industrial Forestry Complex

Cellulose paper production dominates the industrial forest output in the state. With eleven factories operating, this sector produces about 800,000 tons of cellulose per year (1977), amounting to about 49% of Brazilian production (APFPC 1978). In general, the producing units are geographically grouped and localized in the east (Figure 2).

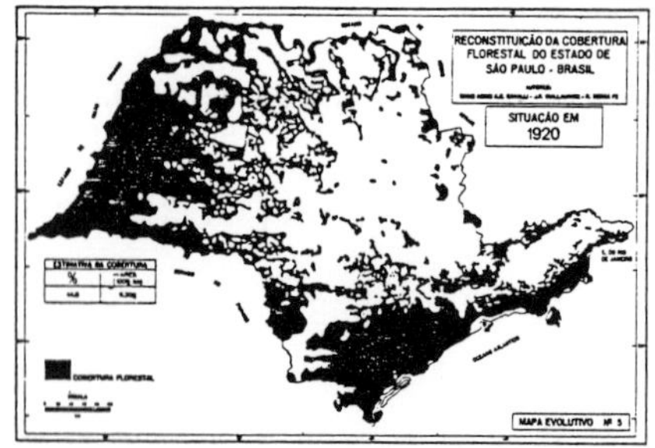

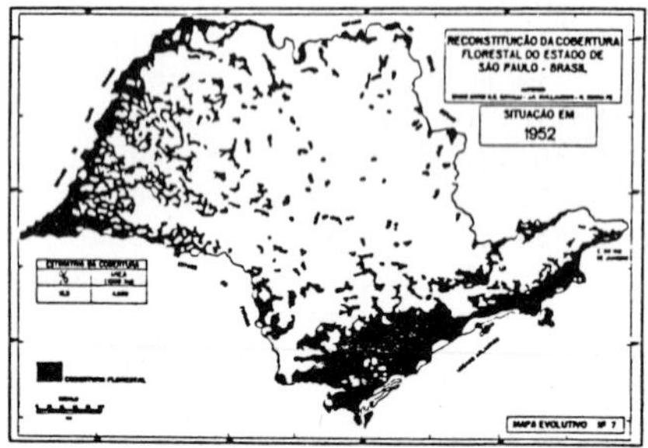

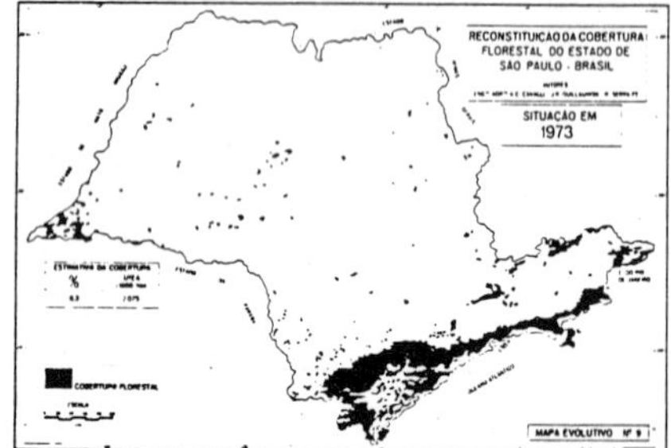

Figure 1. Deforestation trend in the state of São Paulo (Source: Victor 1975).

Table 1. Land areas in the state of São Paulo.

Land Characteristic	Area (ha)	Proportion of Total Land (%)
Native forest	2,069,920	8.33
Brushwood	1,241,090	4.99
"Cerradão" (extensive barren land)	105,390	0.42
"Cerrado" (barren land)	784,990	3.16
"Campo cerrado" (barren plains)	148,390	0.60
Grassland	48,870	0.18
Afforestation	841,420	2.58

Fiberboard is the second most important industrial forest product in the state (Figure 2). Although there are only two factories, their output (600,000 tons in 1978) is more than sufficient for domestic needs, allowing a substantial surplus for export. There are also particleboard factories in São Paulo, whose production in 1973 reached 55 million m^3, or 20% of the national output (Muthoo 1977). There are still about 675 registered sawmills, mostly small and medium size,

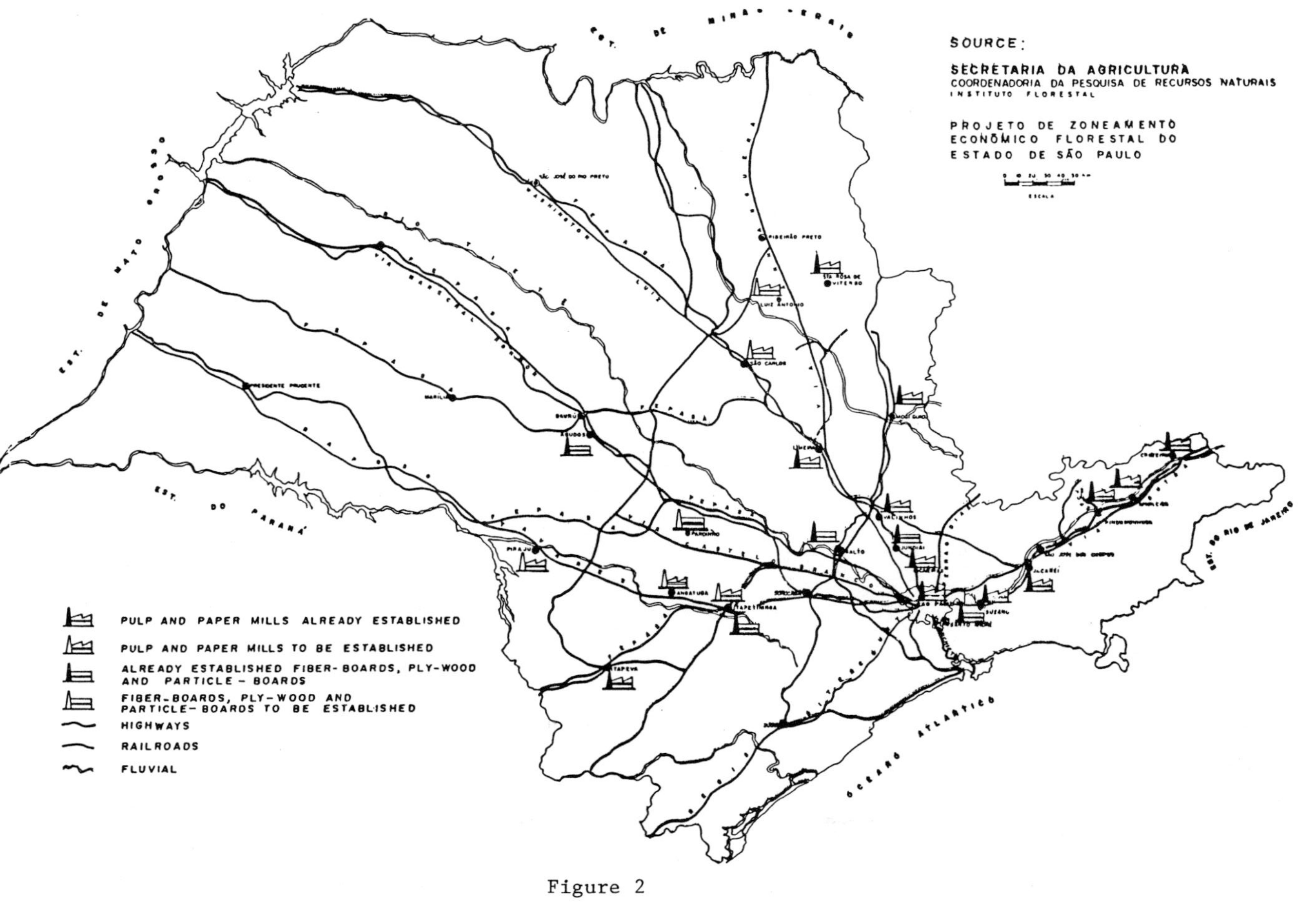
SOURCE:
SECRETARIA DA AGRICULTURA
COORDENADORIA DA PESQUISA DE RECURSOS NATURAIS
INSTITUTO FLORESTAL
PROJETO DE ZONEAMENTO ECONÔMICO FLORESTAL DO ESTADO DE SÃO PAULO
PULP AND PAPER MILLS ALREADY ESTABLISHED
PULP AND PAPER MILLS TO BE ESTABLISHED
ALREADY ESTABLISHED FIBER-BOARDS, PLY-WOOD AND PARTICLE-BOARDS
FIBER-BOARDS, PLY-WOOD AND PARTICLE-BOARDS TO BE ESTABLISHED
HIGHWAYS
RAILROADS
FLUVIAL
EST. DE MINAS GERAIS
EST. DE MATO GROSSO
EST. DO PARANÁ
EST. DO RIO DE JANEIRO
OCEANO ATLANTICO

Figure 2

producing from 2 to 64 m^3 per day, using chiefly hardwood of native species, delivered by other states.

The use of wood as a source of energy is of great importance in the composition of forestry consumption. Estimates of this consumption show values of 4.1 million in 1977 and 4.6 million in 1978 (IBDF/DIC 1978, 1979).

Since the primitive forests of native species are practically exhausted, and the government through strict legislation and constant vigilance discourages their exploitation, the only way to supply the forest industry is through man-made forests. The dominant species are *Pinus* and *Eucalyptus*.

ENVIRONMENTAL PARAMETERS

Tree Species

An essential factor for this study is the correlation between species and site. Investigation of the growth of pines in São Paulo is the first approach, since most of the plantations are no older than fifteen years, which means that only a part of the rotation has been analyzed. Of the pines under observation, trustworthy results for *Pinus elliottii* var. *elliottii* are available, and to a lesser extent for *P. taeda*, *P. patula*, *P. khesya*, *P. caribaea* var. *caribaea*, *P. caribaea* var. *hondurensis*, *P. oocarpa*, and *P. caribaea* var. *bahamensis*.

Earlier studies (Van Goor 1965; Golfari 1967) established remarkable correlations between the growth of *P. elliottii* var. *elliottii* and some components of the site, such as climate, soil (at the level of great group, as an indicator of natural fertility), and former vegetation. Its yield table, however, is a provisional one.

For *Eucalyptus* the situation is reversed: the culture was established in the state more than seventy years ago, and therefore the existing yield tables are sufficiently trustworthy and exhaustively tested. In practice, however, more studies on the correlation between development and site will be necessary (Van Goor 1975). The *Eucalyptus* species generally used are *E. grandis*, *E. saligna*, *E. alba* (hybrid or *urophylla*), *E. urophylla*, *E. citriodora*, *E. camaldulensis*, and *E. tereticornis*.

Permanent Sample Plots

To obtain exact data on the relation between productivity and site, a number of permanent sample plots were established, representing the most diverse edapho-climatic conditions in the state. For *P. elliottii* var. *elliottii* there are 684 plots, selected in accordance with preestablished criteria (Van Goor 1975). Each plot at the end of rotation must contain at least twenty useful trees, and, whenever possible, uncontrollable "external" influences, such as unfamiliarity with the origin of the basic genetic material, must be avoided. A seed of unknown origin can affect productivity to such an extent that the factors of the site may become "disguised."

Site and the Factors of Growth

The factors of site, fundamental for growth, are related to climate and soil. The topography in subtropical and tropical regions does not influence growth directly, for it is implicitly considered in the development of the soil. Growth is, however, directly influenced by the soil.

Climate

In accordance with Thornthwaite, the state of São Paulo can be divided into climatic regions based on the average annual temperature and rain deficit as follows:

Region A: Below 20°C, without hydric deficit.
Region B: Below 20°C, with hydric deficit less than 30 mm.
Region B_1: Below 20°C, with hydric deficit between 30 and 60 mm.
Region C: Above 20°C, without hydric deficit.
Region D: Between 20 and 22°C, with hydric deficit less than 30 mm.
Region E: Between 20 and 22°C, with hydric deficit between 30 and 60 mm.
Region E_1: Between 20 and 22°C, with hydric deficit above 60 mm.
Region F: Above 22°C, with hydric deficit above 60 mm.
Region F_1: Above 22°C, with hydric deficit between 30 and 60 mm.
Region F_{11}: Above 22°C, with hydric deficit below 30 mm.

The definition of these regions is based on water-holding capacity of 300 mm. The regions indicated with the digits 1 and 11 are of lesser importance and in this study were included in the principal types of climate as shown in Figure 3.

Based on essentially climatic parameters, the state of São Paulo was classified in regions qualified and not qualified for the cultivation of several kinds of *Pinus* (Golfari 1967) and later several kinds of *Eucalyptus*.

Soils

Soil factors are divided into physical, chemical, and biological. Physical factors determine the conditions for nitrogen fixing, mycorrhizae, and so forth, and are well defined by the great soil groups. The fertility of the soil is more difficult to determine. Since the soil is covered with natural vegetation, the fertility is reasonably related to the great soil groups. But even under natural conditions, fertility varies. As one discovers in practice, there is variation of natural vegetation on the same unit of soil; the atropical influence is so remarkable that the fertility of the site can be determined only through a chemical analysis.

Concerning physical properties, particular soils encountered include latosols, podzolized soils, lithosols, hydromorphic soils, regosols, and Mediterranean soils. From the great soil group in the state of São Paulo, Pv (podzolized red green), PVp (podzolized red green variation Piracicaba), Plm (podzolized variation Lins), and Pml (podzolized variation Marília) were grouped.

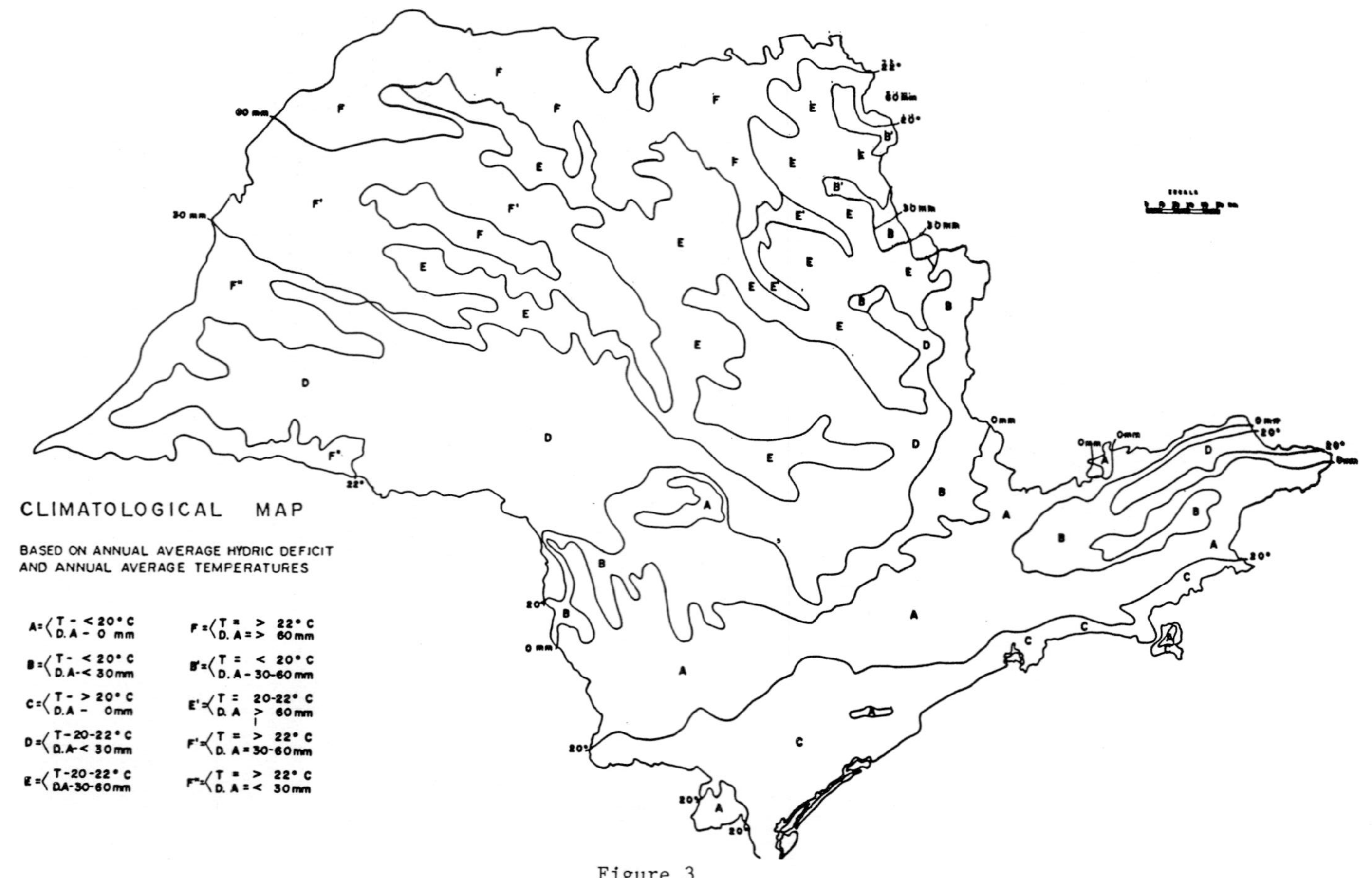
CLIMATOLOGICAL MAP
BASED ON ANNUAL AVERAGE HYDRIC DEFICIT
AND ANNUAL AVERAGE TEMPERATURES
A = T - < 20° C; D.A - 0 mm
B = T - < 20° C; D.A - < 30 mm
C = T - > 20° C; D.A - 0 mm
D = T - 20-22° C; D.A - < 30 mm
E = T - 20-22° C; D.A - 30-60 mm
F = T = > 22° C; D. A = > 60 mm
B' = T = < 20° C; D. A - 30-60 mm
E' = T = 20-22° C; D. A > 60 mm
F' = T = > 22° C; D. A = 30-60 mm
F" = T = > 22° C; D. A = < 30 mm

Figure 3

Adding climatic parameters to edaphic ones, one can gain good information on the volume of wood to be expected in a certain region (Van Goor 1975). Figure 4 gives details.

Site as an Ecological Unit

Ecological forest units are defined by climate and great soil groups. An additional criterion is the vegetation. Whenever possible, permanent sample lots representative of the ecological forest unit have been established on sites. Ecological units composed of soils of high fertility and therefore of high value to agriculture were not included, such as terra rossa soil, podzolized soils from Lins and Marília, and red-yellow Mediterranean soil. For the same reason, units of hydromorphic soil and lithosols were not included because they are not suitable for extensive forestry.

Management Systems and Yield Tables

Microregions were the key to economic stratification. In every homogenous microregion several situations were simulated as shown in Table 2. These simulations considered several species (*Pinus* and *Eucalyptus*) under different regimes (simple coppice and high forest) and the respective level of wood output.

For *Eucalyptus* the "regime" of simple coppice in short rotation is already a classic in the state of São Paulo. The raw material resulting from this procedure is traditionally used in pulp and paper mills, for fiberboard, in charcoal manufacture, and so forth. However, the regime of high forest, with longer rotation and trees of larger diameter for sawmills, is not yet established. This reflects the tendency to adapt systems of management used in South Africa (Ramos 1973). The same is true of *Pinus* species, for which models of management systems used abroad were adapted.

It becomes clear that existing plantations in the country already are providing fundamental information for projected future production (Kronka et al. 1971). This is a field in which investigation is being intensified and existing data are provisional.

Since the species of *Pinus* show great differences in development, they have been grouped in two distinct classes: (1) temperate *Pinus*, which tend to develop slowly, include *Pinus elliottii* var. *elliottii*, *P. taeda*, *P. patula*, *P. khesya* and *P. elliottii* var. *densa*; (2) tropical *Pinus*, which develop much faster, include *P. caribaea* var. *caribaea*, *P. caribaea* var. *hondurensis*, *P. caribaea* var. *bahamensis*, and *P. oocarpa*.

Tables 3, 4, 5, and 6 show the yield tables for temperate *Pinus* and *Eucalyptus* species with respectively short and long rotations.

ECONOMIC PARAMETERS

In order to analyze the suitability of different regions of the state for afforestation, one uses the potential profitability coefficient to obtain for each region the level of profitability that can be reached by production in accordance with the species and the rotation analyzed. This also reveals the rate of return on investment. Later one makes a separate analysis of transportation conditions and the possibilities of marketing the raw material.

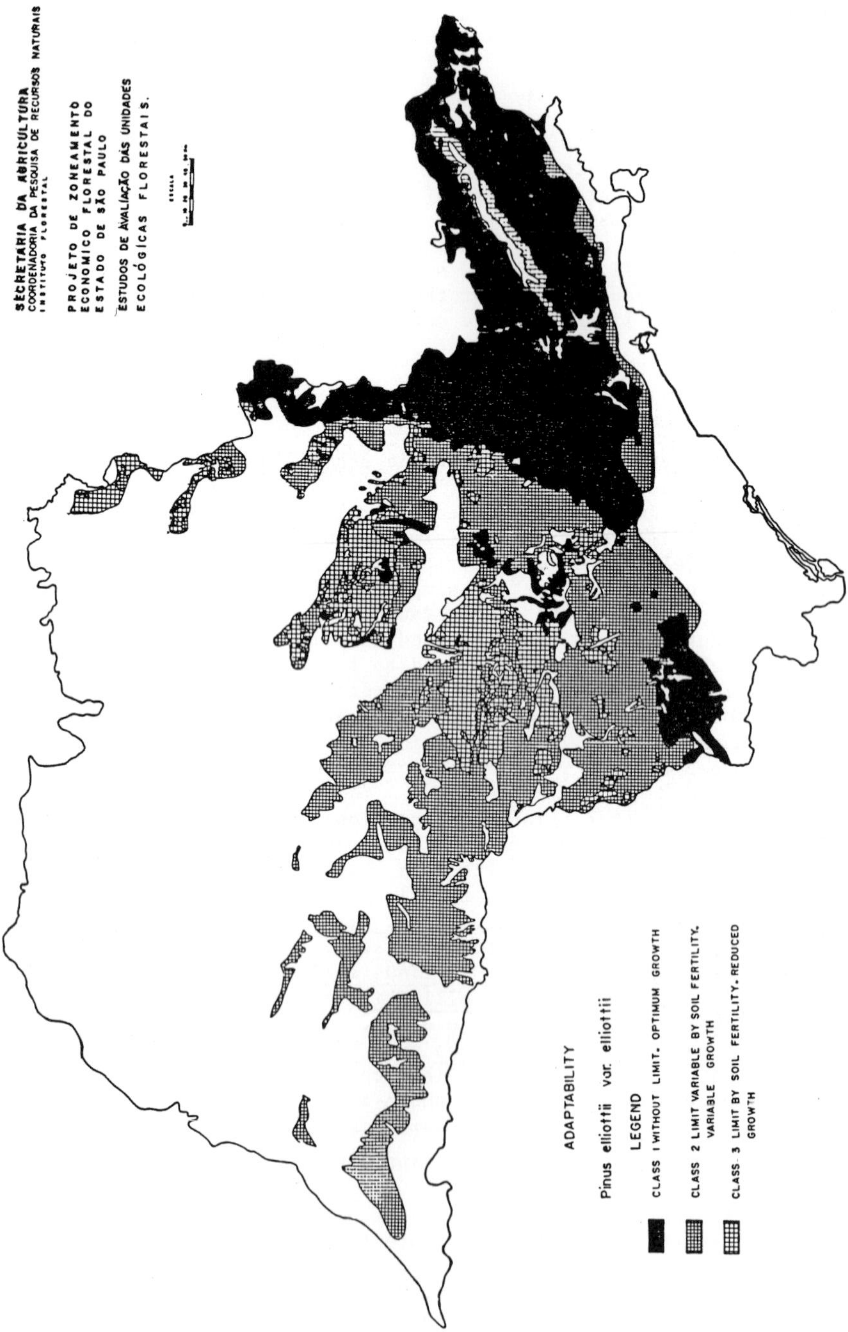

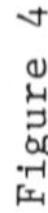
SECRETARIA DA AGRICULTURA
COORDENADORIA DA PESQUISA DE RECURSOS NATURAIS
INSTITUTO FLORESTAL
PROJETO DE ZONEAMENTO
ECONOMICO FLORESTAL DO
ESTADO DE SÃO PAULO
ESTUDOS DE AVALIAÇÃO DAS UNIDADES
ECOLÓGICAS FLORESTAIS.
ADAPTABILITY
Pinus elliottii var. elliottii
LEGEND
CLASS 1 WITHOUT LIMIT. OPTIMUM GROWTH
CLASS 2 LIMIT VARIABLE BY SOIL FERTILITY. VARIABLE GROWTH
CLASS 3 LIMIT BY SOIL FERTILITY. REDUCED GROWTH

Figure 4

Table 2. Tree species according to regime and rotation.

Species	Regime	Rotation (yrs)	Yield Table
Eucalyptus	simple coppice	short (17)	P1 P2 P3 P4
	high forest	short (20)	P1 P2 P3 P4
	high forest	long (35)	P1 P2 P3 P4
Tropical Pinus	high forest	short (20)	P1 P2 P3
	high forest	long (35)	P1 P2 P3
Temperate Pinus	high forest	short (20)	P1 P2 P3
	high forest	long (35)	P1 P2 P3

The coefficient of potential profitability is calculated using the basic forest equation of Hundeshagen. From the relation between revenue and invested capital one determines the internal profitability of the investments in the form of an annual interest rate. This comparison of revenue with cost is done at the same time. The total of all revenues adjusted to the beginning of rotation is:

$$\Sigma R = \left(\frac{Da}{1{,}0\ i^{a}} + \frac{Db}{1{,}0\ i^{b}} + \cdots\cdots\cdots \frac{Cr}{1{,}0\ i^{r}}\right) \quad (P-E)$$

Table 3. Yields for Eucalyptus species (simple coppice).*

	Age (yrs)	Outside Bark (m^3)	Inside Bark (m^3)	Stacked Volume Inside Bark (m^3)
P1	7	375	300	360
	12	161	129	155
	17	128	102	122
	Total	664	531	637
P2	7	285	228	273
	12	122	98	118
	17	97	77	92
	Total	504	403	483
P3	7	214	171	205
	12	91	73	87
	17	73	58	70
	Total	378	302	362
P4	7	153	122	146
	12	65	52	62
	17	52	42	51
	Total	270	216	259

*Percentage of bark = 20%. Bark factor = 0.894. Form factor = 0.55. Coefficient of stacking = 1.2. Square bark factor = 0.80 (% of wood).

The total of costs adjusted to the beginning of rotation is:

$$\Sigma C = PL + \frac{S_1}{1,0\ i^1} + \frac{S_2}{1,0\ i^2} + \frac{a\ (1,0\ i^r - 1)}{0,0i \cdot 1,0\ i^r} + \frac{T\ (1,0\ i^r - 1)}{1,0\ i^r}$$

where:

Da, Db, etc. = amount of thinned wood during the years a, b, and so on
Cr = amount of wood at final cutting
P = price of wood at factory
E = cost of exploitation (cutting, debarking, piling)
PL = cost of planting and silvic treatment during the first year
S_1, S_2 = cost of silvic treatment during the following years
a = annual cost of management, including maintenance and depreciation of installations
T = land value

Table 4. Yields for Eucalyptus species (rotation 35 years).*

	Age (yrs)	Outside Bark Thinned (m^3)	Inside Bark Thinned (m^3)	Inside Bark Thinned, as Pulp (m^3)	Stacked Volume Inside Bark, as Pulp (m^3)	Inside Bark for Sawmill (m^3)
P1	7	105	84	84	101	--
	9	128	102	77	92	25
	12	156	124	56	67	68
	16	175	141	22	26	119
	23	198	158	16	19	142
	35	774	619	62	75	557
	Total	1,536	1,228	317	380	911
P2	7	80	64	64	77	--
	9	97	77	70	84	7
	12	113	91	46	55	45
	16	127	102	21	25	81
	23	141	113	12	14	101
	35	568	454	54	65	400
	Total	1,126	901	267	320	634
P3	7	60	48	48	58	--
	9	72	57	53	63	4
	12	84	67	44	53	23
	16	94	75	15	18	60
	23	103	83	10	13	73
	35	412	330	33	38	297
	Total	825	660	203	243	457
P4	7	43	34	34	41	--
	9	51	41	41	49	--
	12	60	48	34	41	14
	16	66	53	16	19	37
	23	73	58	9	11	49
	35	290	232	24	29	208
	Total	583	466	158	190	308

*See note for Table 3.

The potental profitability coefficient (i) is determined by comparing

$$\Sigma_r = \Sigma_c.$$

To calculate the economic transportation radius of wood from the site of production (plantation) to the place of consumption (industry), one again uses the basic forest equation. It gives the total available for transportation (T_r), once the profitability rate demanded for the investment is fixed. The result is obtained by transforming the basic forest equations:

$$P - E - T_r = \frac{\Sigma c}{\Sigma q}$$

Table 5. Yields for temperate Pinus (rotation 35 years, spacing 2 x 2 m).*

	Age (yrs)	Outside Bark Thinned (m^3)	Inside Bark Thinned (m^3)	Inside Bark Thinned, as Pulp (m^3)	Stacked Volume Inside Bark, as Pulp (m^3)	Inside Bark for Sawmill (m^3)
P1	7	46	34	34	44	--
	9	56	42	32	41	10
	12	72	54	24	31	30
	16	91	68	10	13	58
	23	105	79	8	10	71
	35	470	353	35	45	318
	Total	840	630	143	184	487
P2	7	35	26	26	34	--
	9	43	32	28	35	4
	12	54	40	22	28	18
	16	69	52	11	15	41
	23	80	60	6	8	54
	35	341	256	26	34	230
	Total	622	466	119	154	347
P3	7	25	19	19	24	--
	9	32	24	24	31	--
	12	40	30	20	26	10
	16	51	38	10	13	28
	23	60	45	5	7	40
	35	256	192	20	26	172
	Total	464	348	98	127	250

*Percentage of bark = 25%. Bark factor = 0.866. Form factor = 0.50. Coefficient of stacking = 1.30. Square bark factor = 0.75 (% of wood).

or

$$T_r = P - E - \frac{\Sigma c}{\Sigma q}$$

dividing T_r by the cost of transportation for m3/km of wood, one determines the maximum possible radius for transportation.

It becomes evident that when the effective rentability is equal to the potential rentability, the sum available for transportation will be equal to zero and the transportation of wood will not be economical.

Determination of Homogenous Microregions

The microregions which supplied the basis for the potential profitability coefficient consist of homogenous areas with the following configuration: natural vegetation, topographic conditions, ecological forest units (which determine the development of the species), and land value. For each microregion, the cost of silvicultural operations for planting and maintaining the plantation were characterized. In the same

Table 6. Yields for temperate *Pinus* (rotation 20 years, spacing 2 x 2 m).*

	Age (yrs)	Outside Bark Thinned (m^3)	Inside Bark Thinned (m^3)	Inside Bark Thinned, as Pulp (m^3)	Stacked Volume Inside Bark, as Pulp (m^3)	Inside Bark for Sawmill (m^3)
P1	7	46	34	34	44	--
	9	56	42	32	41	10
	12	72	54	24	31	30
	16	91	68	10	13	58
	20	325	244	25	33	219
	Total	590	442	125	162	317
P2	7	35	26	26	34	--
	9	43	32	28	35	4
	12	54	40	22	28	18
	16	69	52	11	15	41
	20	232	174	18	25	156
	Total	433	324	105	137	219
P3	7	25	19	19	24	--
	9	32	24	24	31	--
	12	40	30	20	26	10
	16	51	38	10	13	28
	20	170	128	13	17	115
	Total	318	239	86	111	153

*See note for Table 5.

way the operational costs of exploitation and the prices for the raw material resulting from this exploitation were fixed.

Silvicultural Operating Costs

Costs for planting, maintenance, and exploitation were derived from the topographic variables, types of existing vegetation, and the species used (*Eucalyptus* and *Pinus*). Planting varies to a great extent with the topography: for mountainous regions it is totally by hand, for wavy topography it is by hand and mechanized, and for plains it is completely mechanized.

The silvicultural operations considered for the year of planting were site preparation, protection, cultural traits, expenses for tools and products, infrastructure, management, surveying, projects, taxes, and general expenses. For the years of maintenance, the following items were considered: cultural traits, protection, tools and products, conservation, depreciation, management, and general expenses.

For exploitation the following activities were considered: cutting of branches, barking, yarding, and piling for transportation. Different costs were arrived at for differences in topography and final destination of the raw material (wood for pulp or sawmills). For transportation, only trucks were considered, based on prices in force in October 1974. A survey of the industries of the sector provided prices for wood at the sawmills as well as for other uses.

Topography, Vegetation, and Value of the Land

In order to obtain the cartographic representation of homogeneous microregions, mapping of topography, vegetation, and land value was necessary (see Figure 5). Lands with slope up to 12% were grouped in the category "wavy plains," those with slope from 12% to 20% as "strongly wavy," those between 20% and 40% as "mountainous," and those above 40% as "steep" (protected areas). As for existing vegetation, basically grassland, brushwood, barren land, and extensive barren land were taken in account (Figure 6).

The survey of land suitable for farmland revealed land values in distinct regions, which also were mapped. The types considered were first-class and second-class culture land, pastures, land for afforestation, and grassland.

Data Processing

Using all data collected, maps were made of homogeneous microregions. For the *Pinus* species, 1,516 homogeneous regions were established. The potential profitability coefficient was calculated for each microregion, executing also a number of simulations on the species (*Pinus* or *Eucalyptus*) and distinct regimes (simple coppice or high forest). Because of the complexity of calculation and the great number of operations involved, electronic processing is fundamental.

Mapping of Conditions for Transport and Profitability

In order to systematize the mapping of the conditions in the microregions, the types of potential profitability are grouped in the following classes: negative (0), minimum (0.1 to 5%), regular (5.1 to 8%), medium (8.1 to 11%), good (11.1 to 14%), and excellent (over 14%).

In accordance with this criterion, Figure 7 was organized showing the potential profitability for the culture of *Pinus* species in the state of São Paulo, subject to the regime of high forest with a rotation of twenty years.

To map transportation conditions, it is necessary to establish the level of internal profitability desired for the investment. For the state of São Paulo, this index was fixed at 8%, a rate widely accepted for investments of this nature.

The conditions for transportation are grouped as follows: negative, 0-50 km, 51-100 km, 101-200 km, and more than 200 km. The cartographic representation of these results is shown on Figure 8.

Delimitation of Priority Areas for Afforestation

For this delimitation, the following data were fundamental: (1) the parameters for conditions for transportation of *Eucalyptus* species under the simple coppice regime (this is justified, since *Eucalyptus* is the traditional species in the state and the greatest source of raw material for many industries); (2) the industries of the sector and the respective transportation network (Figure 2); and (3) critical areas that demand a greater degree of conservation, and therefore were not included in priority areas for industrial afforestation, such as the Vale do Ribeira.

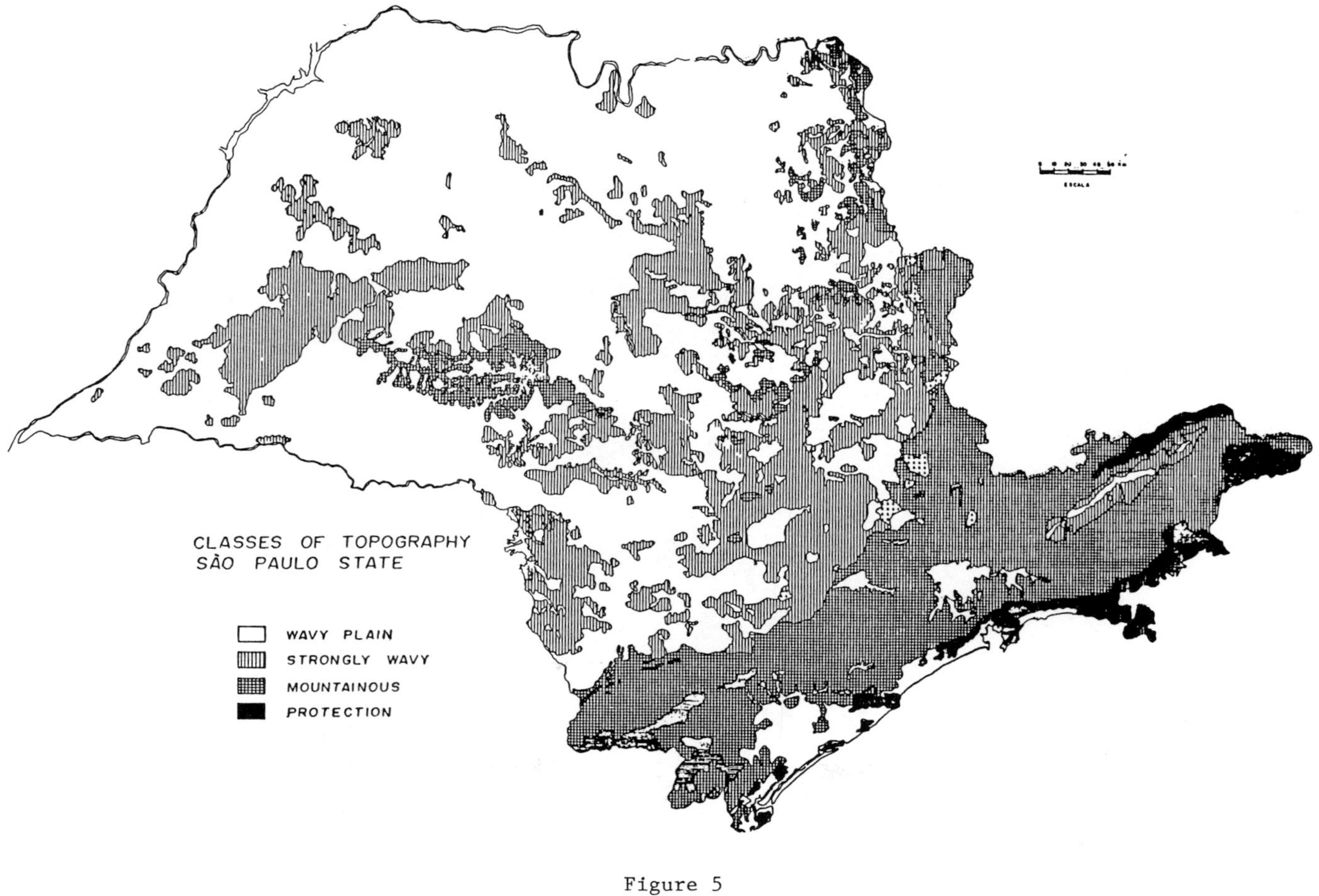
CLASSES OF TOPOGRAPHY
SÃO PAULO STATE
WAVY PLAIN
STRONGLY WAVY
MOUNTAINOUS
PROTECTION

Figure 5

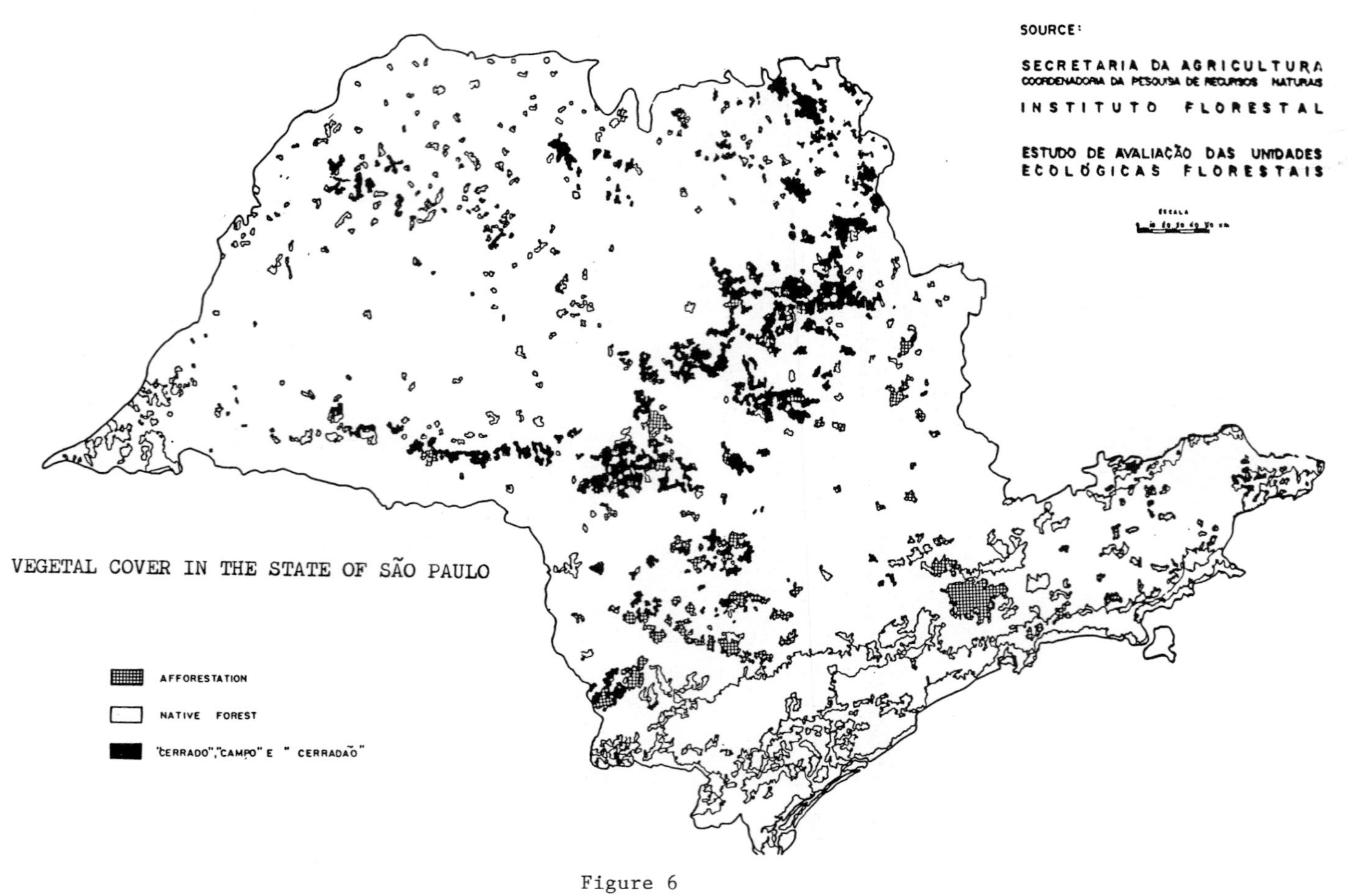
SOURCE:
SECRETARIA DA AGRICULTURA
COORDENADORIA DA PESQUISA DE RECURSOS NATURAIS
INSTITUTO FLORESTAL
ESTUDO DE AVALIAÇÃO DAS UNIDADES
ECOLÓGICAS FLORESTAIS
VEGETAL COVER IN THE STATE OF SÃO PAULO
AFFORESTATION
NATIVE FOREST
"CERRADO", "CAMPO" E "CERRADÃO"

Figure 6

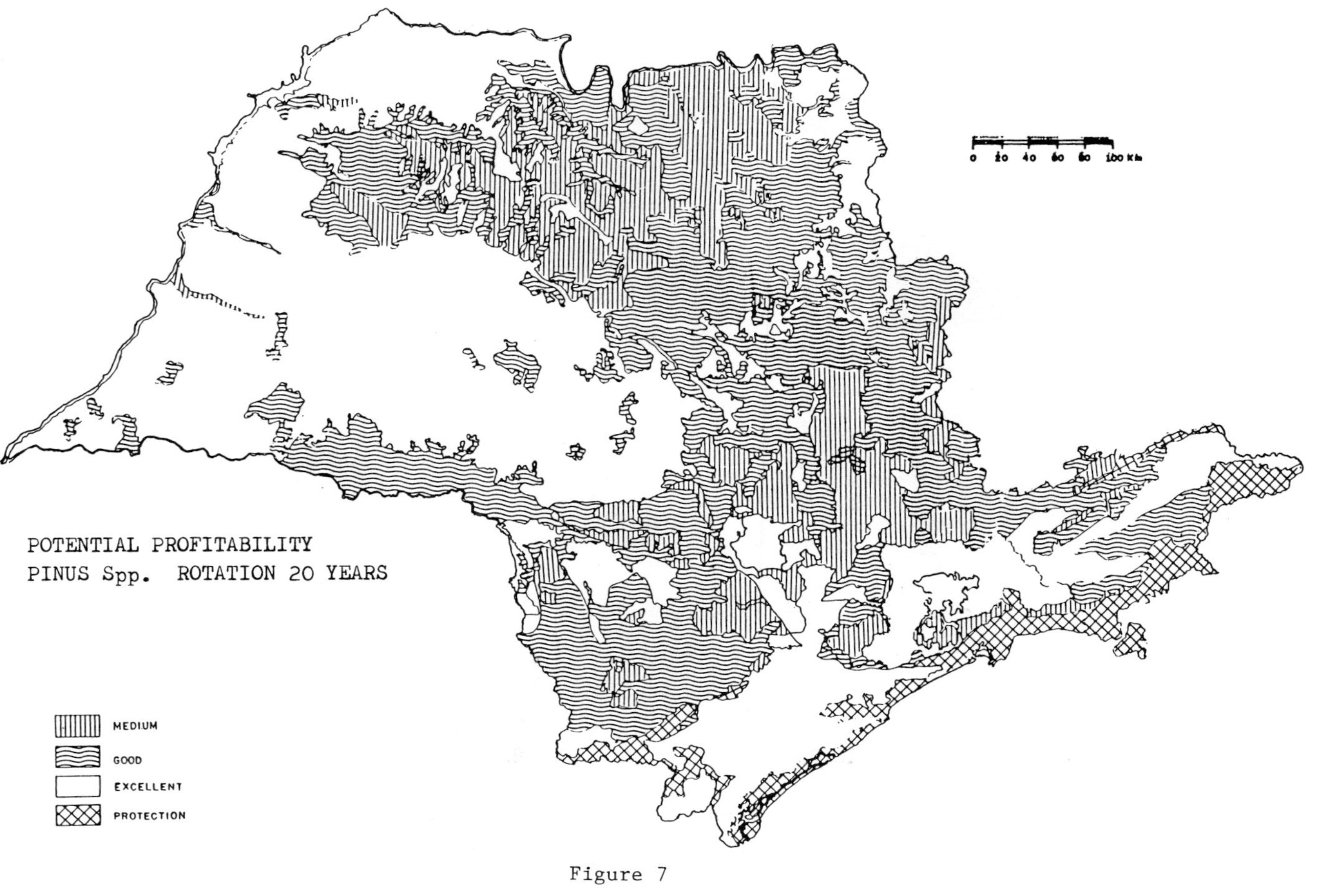
0 20 40 60 80 100 Km
POTENTIAL PROFITABILITY
PINUS Spp. ROTATION 20 YEARS
MEDIUM
GOOD
EXCELLENT
PROTECTION

Figure 7

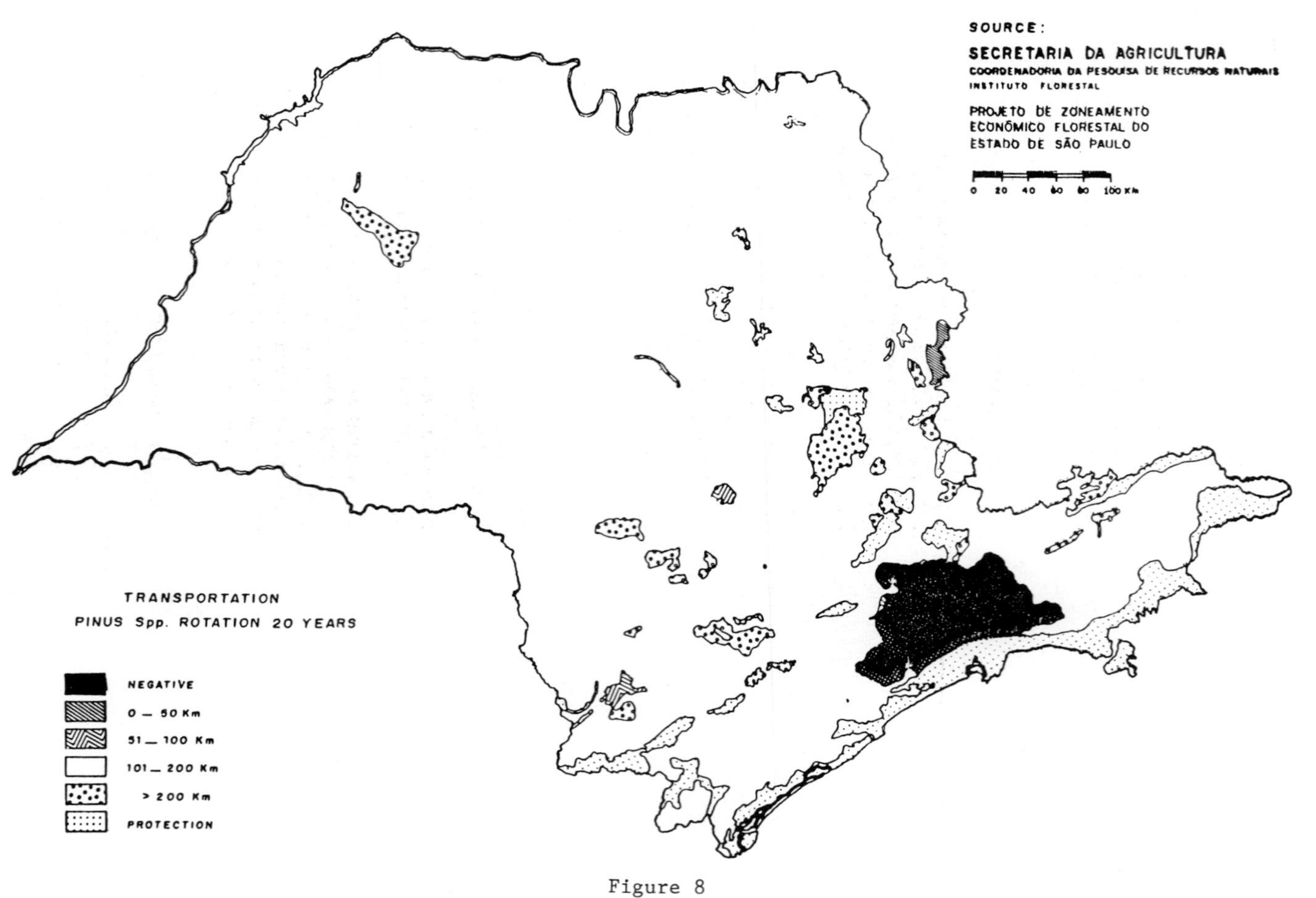
SOURCE:
SECRETARIA DA AGRICULTURA
COORDENADORIA DA PESQUISA DE RECURSOS NATURAIS
INSTITUTO FLORESTAL
PROJETO DE ZONEAMENTO
ECONÔMICO FLORESTAL DO
ESTADO DE SÃO PAULO
0 20 40 60 80 100 Km
TRANSPORTATION
PINUS Spp. ROTATION 20 YEARS
NEGATIVE
0 — 50 Km
51 — 100 Km
101 — 200 Km
> 200 Km
PROTECTION

Figure 8

The cartographic representation of these priority areas is shown in Figure 9. As one can see, practically half the state was included in the priority area. In practice, subsidies are given only to approved projects in these areas.

Since the advent of the law for fiscal subsidies, practically 600,000 hectares of forests have been planted (Figure 10), so that today the state has about 800,000 hectares of *Eucalyptus* and *Pinus* forests, predominantly the first species over the second.

The spacial distribution of this afforestation can be seen on Figure 11, which is based on aerial photographs taken in 1973-74. This figure allows the conclusion that the orientation proposed by successive zoning is being attended to.

DISCUSSION AND CONCLUSION

The ecological and economic zoning that was established in the state of São Paulo in 1966 became a powerful instrument in the hands of the public administration to orient and discipline industrial afforestation, executed by private enterprise. However, an appraisal of the period shows some distortion, conceptual as well as technical, which must be corrected in due time, once zoning is understood as a dynamic study.

These distortions may be summarized as follows:

1. The preliminary zoning did not give due emphasis to the use of wood as a source of energy for small, medium, and large industries. As a consequence of the acute energy crisis, it was seen that the state has a significant potential to substitute the traditional fossil energy for that of biomass, especially of forest origin. This problem is not confined to priority zones for afforestation but exists in the state as a whole.

2. The permanent sample plots must perform as such in order to give consistency to production tables and a better knowledge of the relation between site and growth, chiefly concerning *Eucalyptus*.

3. Genetic tree improvement, heavy fertilization, and intensive management are leading to an impressive increase in volume of the species employed. This is shown in Table 7. It is evident that these data will alter the relation between cost and revenue.

4. In the same way, the costs for transportation will be affected by prices of petroleum and derivates, and therefore the original limits of priority areas must be reviewed.

5. In the economic model used, the remuneration of the factor land is given through an annual rent. It is clear that rent is in proportion to the price of the land: the higher the value, the higher the rent. Recently the price of land suffered an abnormal increase, owing to a series of atypical considerations, such as speculation on real estate. As a result, afforestation becomes practically impossible on a

Table 7. Trends in biomass production (mean annual increment, m^3/ha/yr, with bark).

Species	1970	1980	1985 (outlook)
Eucalyptus	20	40	50
Pinus	18	35	43

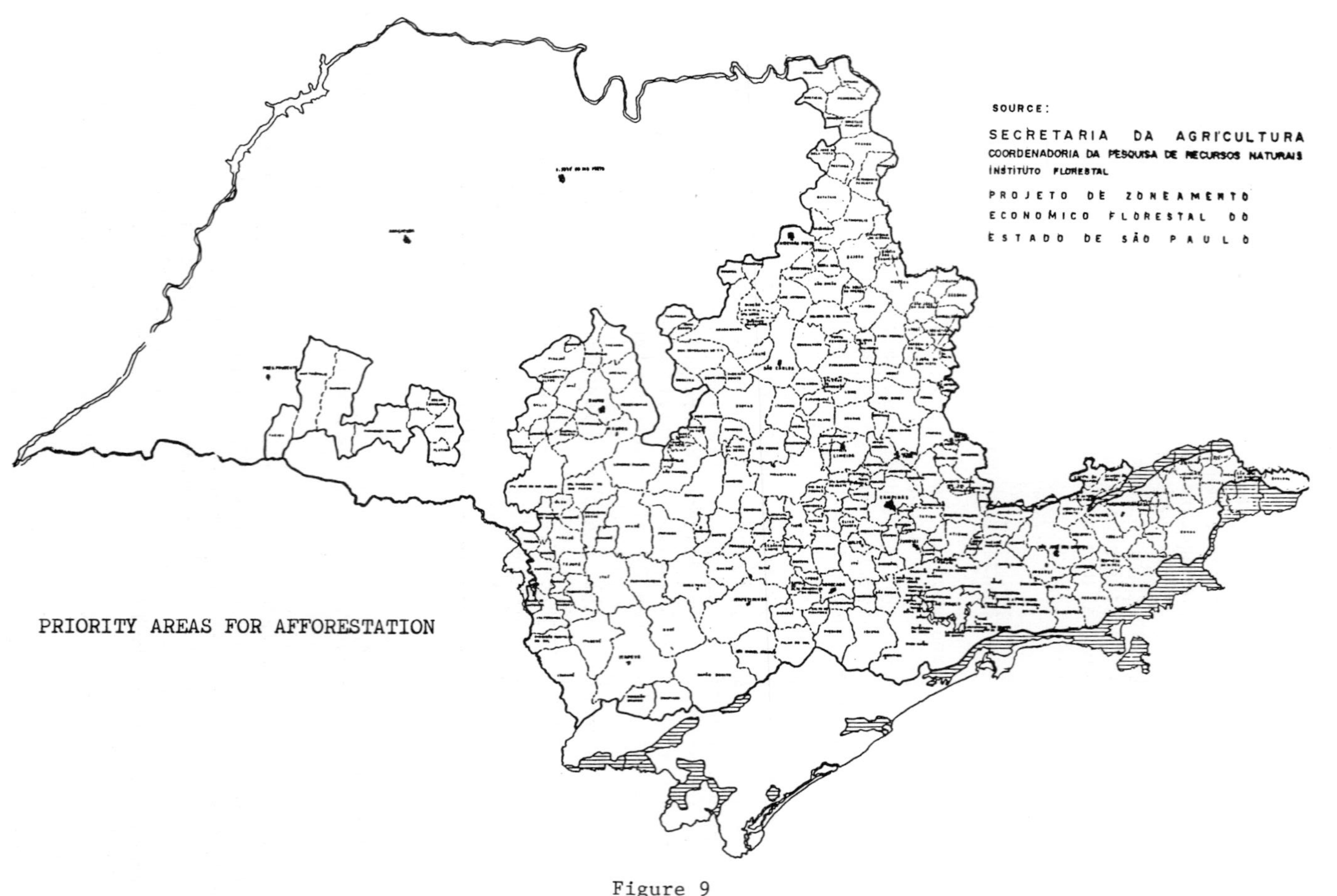

PRIORITY AREAS FOR AFFORESTATION

Figure 9

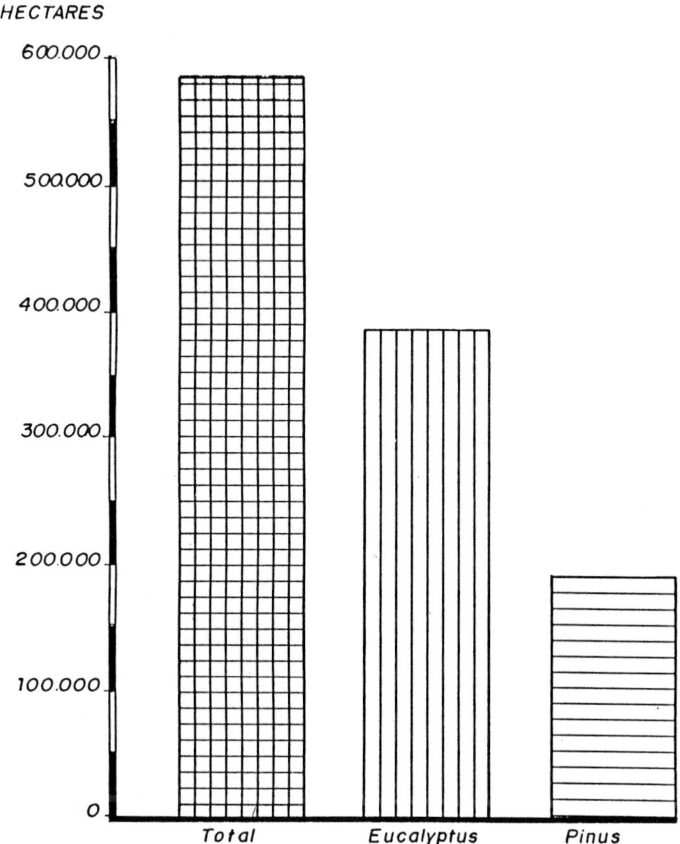

Figure 10. Subsidized afforested areas of the state of São Paulo, 1967-79.

substantial part of the territory. It is therefore necessary that the model consider another form of remuneration for this factor of production, more in accordance with reality in São Paulo.

6. In the adopted system, subsidized afforestation has favored only large-scale investments, to the detriment of small and medium ones, thus jeopardizing the necessary equilibrium of supply to the industry. In part, this distortion has been corrected with the implementation of the Program of Energetic Forestry for Afforestation of Small and Medium Farms.

7. It is fundamental that industrial afforestation must be in harmony with industrial zoning in the state, so that the transportation industry, well known for a high level of pollution, may not contaminate the preserved ecological systems.

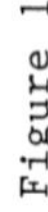

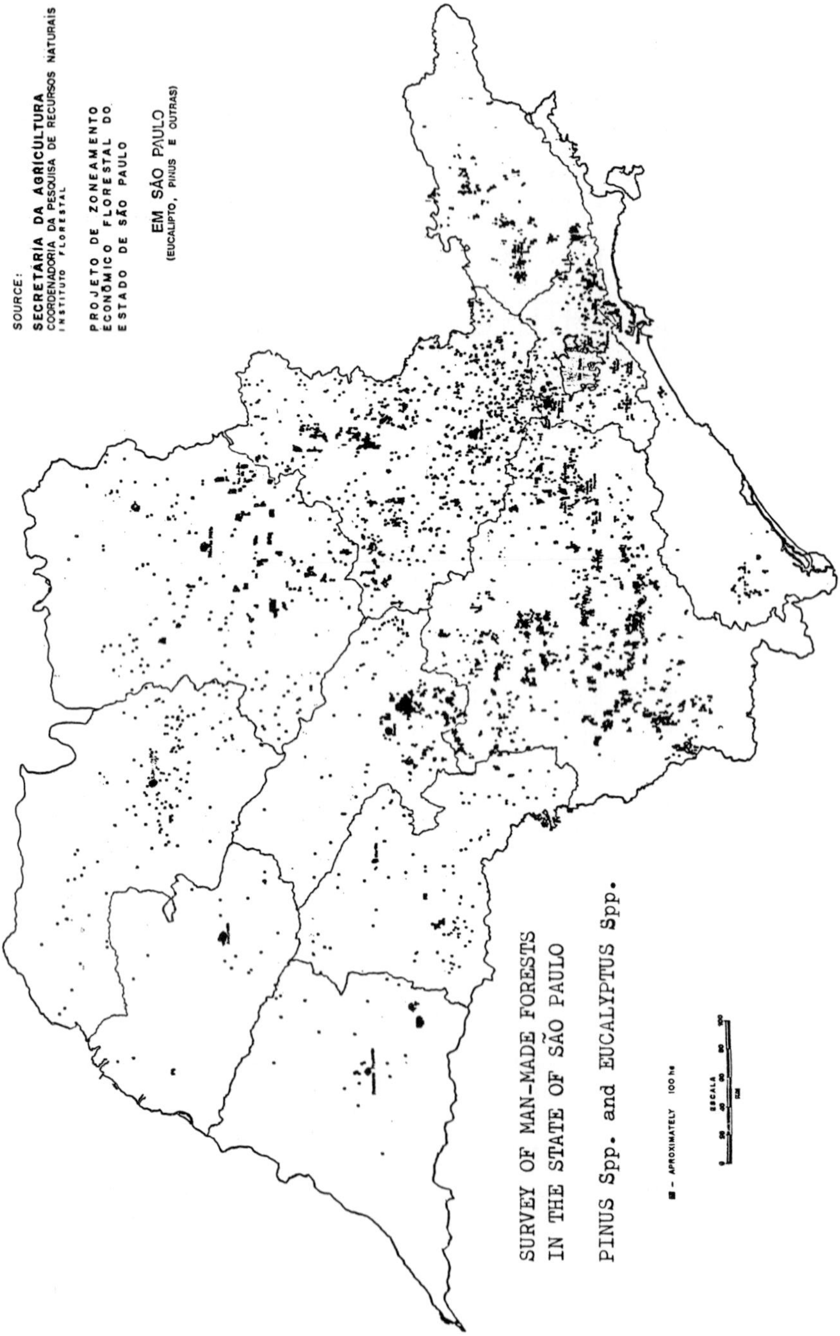
SOURCE:
SECRETÁRIA DA AGRICULTURA
COORDENADORIA DA PESQUISA DE RECURSOS NATURAIS
INSTITUTO FLORESTAL
PROJETO DE ZONEAMENTO
ECONÔMICO FLORESTAL DO
ESTADO DE SÃO PAULO
EM SÃO PAULO
(EUCALIPTO, PINUS E OUTRAS)
SURVEY OF MAN-MADE FORESTS
IN THE STATE OF SÃO PAULO
PINUS Spp. and EUCALYPTUS Spp.
- APROXIMATELY 100 ha
ESCALA
0 20 40 60 80 100

Figure 11

REFERENCES

Associação Paulista dos Fabricantes de Papel e Celulose (APFPC). 1978. Relatório estatístico 1977. São Paulo.

IBDF/DIC. 1978. Quadro demonstrativo do consumo de lenha e carvão vegetal pelas cerâmicas e olarias no Brasil em 1977. Brasília. 2 pp.

__________. 1979. Quadro demonstrativo do consumo de lenha e carvão vegetal pelas cerâmicas e olarias no Brasil em 1978. Brasília. 2 pp.

Golfari, L. 1967. Coníferas aptas para repoblaciones forestales en el estado de São Paulo. Silvicultura em São Paulo 6:7-62.

Kronka, F. J. N. et al. 1971. Dados informativos sobre desbastes executados em plantações de *Pinus* spp no estado de São Paulo. Associação Técnica Brasileira de Celulose e Papel. Tese apresentada a IV Convenção Anual. Semana do Papel. São Paulo. 10 pp.

Muthoo, M. K. 1977. Perspectivas e tendências do setor florestal Brasileiro, 1975-2000. Desenvolvimento e Planejamento Florestal. Série Técnica, Brasília (4).

Ramos, I. 1973. África do Sul, horizonte florestal do Brasil: O eucalipto, madeira de serraria na África do Sul. Edit. Gomes, São Paulo. 81 pp.

Secretaria da Agricultura. 1970. Programa florestal do estado de São Paulo. São Paulo. 65 pp.

__________. 1975. Zoneamento econômico florestal do estado de São Paulo. Instituto Florestal. Bol. Técnico 17. São Paulo. 80 pp.

Van Goor, C. P. 1965. Reflorestamento com coníferas no Brasil. Dep. Rec. Nat. Ren. Ministério da Agricultura. Bol. 9. 58 pp.

__________. 1975. Crescimento de *P. elliottii*: Unidades ecológicas florestais e pesquisa florestal. Instituto Florestal. Bol. Técnico 14. São Paulo. 50 pp.

Ventura, A. 1964. Problemas técnicos da silvicultura paulista. Silvicultura em São Paulo 3:61-80.

Victor, M. A. M. 1975. A devastação florestal. Sociedade Brasileira de Silvicultura, São Paulo.

8

METHODOLOGICAL ASSUMPTIONS FOR FOREST SOIL AND SITE CHARACTERIZATION IN THE KAL-BIGLEB SUBSYSTEM

A. Kowalkowski

INTRODUCTION

The importance of forests in the human environment is disproportionate to the space they occupy. There has recently been a great increase in attention to beneficial environment-shaping effects in the anthropocentric forest landscape. Consequently, it is understandable that there is a growing interest in all problems concerning forests of an economic and social nature, in an effort to optimize the use of the natural space. This optimization, however, will only be possible after there has been an appropriate systematization and synthetizing of the data that form the elementary constituents of forest ecosystems.

The forest is a geographical phenomenon and as such ought to be considered from the angle of the planetary, landscape, and chorological axioms of the natural space (Neef 1967). In order to define the inventory of ecosystem constituents, as well as the development stage of the forest ecosystem, the following factors have been assumed as a basis: (1) topological differentiation of the constituent elements, (2) chorological differentiation, separation, and spatial combinations of the constituents, and (3) homological relations as well as the place of the constituents according to development stage, age, and maturity.

With regard to causes (Kurkin 1973; Neef 1967), the above factors can be divided into entopical (climate, geological structure), ecotopical (water conditions of soils and nutritional conditions), and biotopical (phytocoenosis, zoocoenosis, and anthropocoenosis). The soils, for instance, develop under the influence of varying entopical, ecotopical, and biotopical factors in the following environments: glacio-hydrogenic, frost- and permafrost ridden, and hydro-, bio-, and anthropogenic (Kopp 1965; Kowalkowski and Borzyszkowski 1977; Kowalkowski 1980). Consequently, in the soil profile, sets of attributes can be found of these environments (Figure 1). Their quantitative and qualitative characteristics define the soil fertility level as well as the productivity of the forest ecosystems.

THE SUBSYSTEM

The KAL-BIGLEB subsystem--Kartografia gleb Leśnych (forest soil cartography), and Bank Informacji o GLEBach (soil data bank)--is complementary to the BIGLEB universal data bank of the soil environment and ecosystems, established by the Polish Soil Science Society (Kowaliński et al. 1979, Figure 2).

The fundamental objective of subsystem KAL-BIGLEB is the collection, storage, processing, transmission, and supply of information and

Environment of the landscape development /after Kopp 1969, Jäger 1979/		Climates	Plant formations	Processes of morphogenetical, geochemical and biological perstructions																																	
				1. Glacial	2. Frost /after Karte 1979/								3. Hydrologic												4. Phyto-zoogenic				5. Anthropogenic							
				01. Translocation	01. Kongelifraction	02. Desquamation	03. Kongeliturbation	04. Kongelicontraction	05. Gelisolifluction	06. Cryoplanation	07. Depergelation	08. Pingo, palsa, etc.	01. Surface water gley	02. Groundwater gley	03. Swamping	04. Denudation	05. Accumulation	06. Decalcification	07. Podzolization	08. Calcification	09. Solodization	10. Solonization	11. Salinization	12. Tacyrization	01. Zoogenic translocation	02. Phytogenic translocation	03. Humus accumulation	04. Peat accumulation	01. Bolization	02. Erosion	03. Thermocarst	04. Soil cultivation	05. Natural resources exploitation	06. Liquid and solid waste	07. Industrial emission	08. Chemization
Glacial		Nival	Glacier	X	O		O							•						•															•	
			Arctic desert	O	X	O	X	•	•	X		O	O	O		X	X	O		O										•	•		•		•	
Periglacial	eu-	Subpolar	Tundra, forest tundra		X	•	X	X	X	X	O	X	X	X	O	•	O	X	O	•	•	•	•	•	•	O	X	X	O	O	X		O	•	•	•
	boreo-	Humid-cool temperate	Taiga		O	•	O	O	O	O	X	O	O	X	X	O	O	X	X	•	O	•	•	•	O	X	X	X	O	X	X	O	X	O	O	O
	xero-	Semiarid-cool temperate	Steppe, meadow-steppe		O	X	•	•	•	•	O	•	•	•	•	X	O	•	•	O	X	X	X	O	X	X	X	•	X	X	O	X	X	X	X	X
Paraperiglacial	boreo-	Humid-warmer temperate	Forest, forest-steppe		•		•							O	O	O		O	O	O					X	X	O	•	O	X		X	X	X	X	X
	xero-	Semiarid-warmer temperate	Semidesert, desert		•	X								•	•	X	X			X		O	X	X	O	•	•		O	O		•	O	•	•	•
Perstruction zones /coding system/			1. Glacial	1101	2101	2102	2103	2104	2105	2106		2108	3101	3102		3104	3105	3106	3107	3108	3109	3110	3111	3112						5102	5103		5105	5106	5107	5108
			2. Periglacial		2201	2202	2203	2204	2205	2206	2207	2208	3201	3202	3203	3204	3205	3206	3207	3208	3209	3210	3211	3212	4201	4202	4203	4204	5201	5202	5203	5204	5205	5206	5207	5208
			3. Paraperiglacial		2301	2302	2303						3301	3302	3303	3304	3305	3306	3307	3308	3309	3310	3311	3312	4301	4302	4303	4304	5301	5302	5303	5304	5305	5306	5307	5308

X Intense perstruction, O moderate perstruction, • weak perstruction.

Figure 1. Schematic chart.

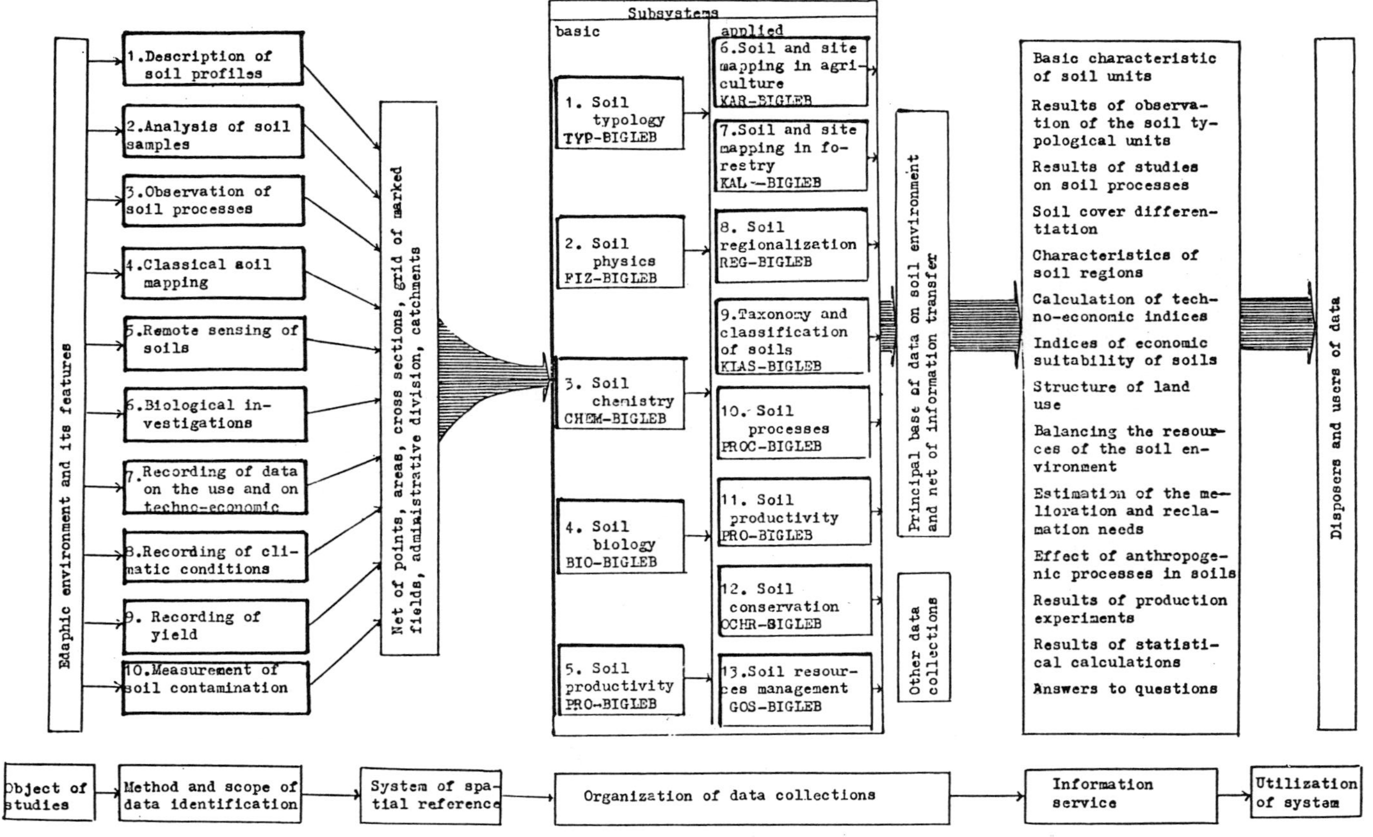

Figure 2. Organizational structure of the BIGLEB information system (after Kowaliński et al. 1979).

syntheses on the environments of the forest soils and ecosystems. Information sets are to create an objective basis for solving economic and scientific problems, such as the following ones:

- descriptions of soil horizon properties (straton, STR), soil profiles (pedon, PED), elementary soil area (ESA), soil combination (SC), elementary forest ecosystems (EFE), types of forest site (TFS), landscape (LS), and administrative units (AU)
- description of the woody, herbaceous, and understory vegetation as regards the items mentioned above: ESA, SC, EFE, TFS, LS, and AU
- assessment of the natural usefulness, size of the required economic operations, and possibility of increasing the soil and site productivity
- assessment of the properties stimulating and limiting the land use according to the objectives assumed
- assessment of the natural and anthropogenic changes in the forest environment
- elaboration of topical map versions of soil, site, agrotechnical operations, natural usefulness, and utilization objectives

The elaboration procedure is presented in Figure 3. The most important geoecological stage of the procedure is to supply appropriate information that characterizes the soil as well as the present and natural sites in order to create objective foundations for defining the natural, present, and future uses of forest ecosystems.

Basic File of the Subsystem

The fundamental assumptions of an efficient data bank are: (1) the files and the file systems may concern only one organizational unit or domain, (2) files must be identical as far as type is concerned, and (3) files are to be formed and serviced with the aid of programs of a single common management system.

The "bank management system" is the programming needed for organizing the data bank. This system ought to make it possible to form the data bank as well as to update and retrieve the data. To ensure the implementation of the assumptions of such a system, the following conditions are necessary: (1) Each card is made up of single, elementary units referred to as document fields; each field, apart from the name ascribed and the relative address, is characterized by length, type of data, and calibrating procedure and characterizes a single attribute being considered. Information is given on the form of storing the data, and the sign of the number. (2) Each document contains eighty vertical and twenty horizontal columns denoting the characteristics and codes of the attributes under study. The sum of all the document fields makes up its content, while the sum of all the cards makes up the information set on the given subject or set of subjects. (3) A definite number of vertical columns is isolated for permanent information, such as file number, topical set number and that of the type of eighty-column card, date the data were obtained, number of the soil profile, and source of the data. (4) Individual fields and sets of fields are provided with the names defined in the heading, as well as with a place for entering verified information. Each separate item of information that defines a given attribute or its constituent part has its field in the vertical columns. (5) Each eighty-column card is provided with explanations as well as with a dictionary, which embodies the index of data and codes.

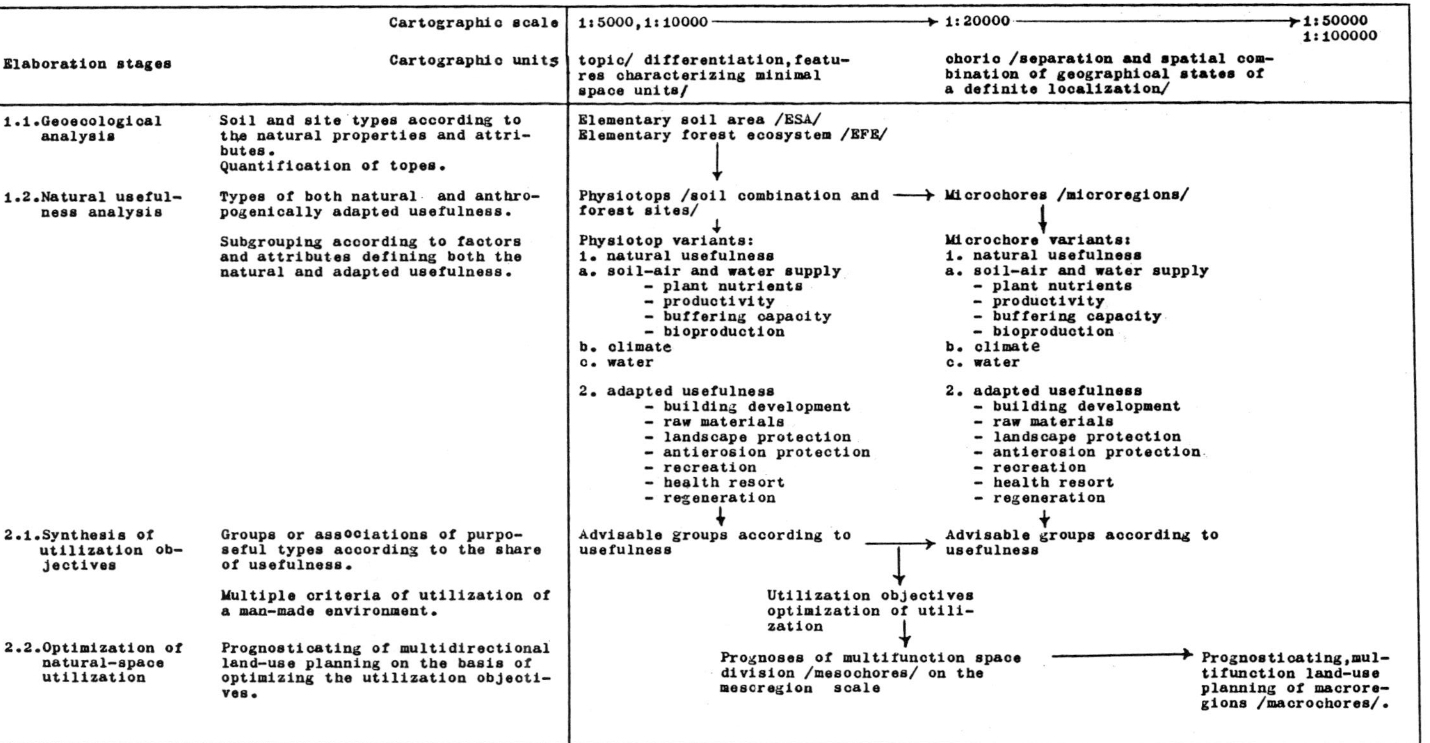

Elaboration stages	Cartographic scale / Cartographic units	1:5000, 1:10000 — topic/ differentiation, features characterizing minimal space units/	1:20000 — choric /separation and spatial combination of geographical states of a definite localization/	1:50000, 1:100000
1.1. Geoecological analysis	Soil and site types according to the natural properties and attributes. Quantification of topes.	Elementary soil area /ESA/ Elementary forest ecosystem /EFE/		
1.2. Natural usefulness analysis	Types of both natural and anthropogenically adapted usefulness. Subgrouping according to factors and attributes defining both the natural and adapted usefulness.	Physiotops /soil combination and forest sites/ Physiotop variants: 1. natural usefulness a. soil-air and water supply - plant nutrients - productivity - buffering capacity - bioproduction b. climate c. water 2. adapted usefulness - building development - raw materials - landscape protection - antierosion protection - recreation - health resort - regeneration	Microchores /microregions/ Microchore variants: 1. natural usefulness a. soil-air and water supply - plant nutrients - productivity - buffering capacity - bioproduction b. climate c. water 2. adapted usefulness - building development - raw materials - landscape protection - antierosion protection - recreation - health resort - regeneration	
2.1. Synthesis of utilization objectives	Groups or associations of purposeful types according to the share of usefulness. Multiple criteria of utilization of a man-made environment.	Advisable groups according to usefulness	Advisable groups according to usefulness Utilization objectives optimization of utilization	
2.2. Optimization of natural-space utilization	Prognosticating of multidirectional land-use planning on the basis of optimizing the utilization objectives.		Prognoses of multifunction space division /mesochores/ on the mescregion scale	Prognosticating, multifunction land-use planning of macroregions /macrochores/.

Figure 3. Organizational conception of studies on soil and forest site mosaics for optimizing the land-use planning of natural space.

Topical Scope of the Basic File

The topical scope of the basic file includes descriptions of the attributes of the natural environment, tree stands, and soils: (1) Cards that describe the environment include localization, regionalization, climate, geological structure, relief, waters, and water conditions. (2) Vegetation and tree stands are described in the cards concerning vegetation, tree stands, chemical composition of the plant material, plant mass, and the like. (3) Soil characterization is entered on cards describing soil profile, physical properties, texture analysis, chemical composition, chemical composition and properties of the humus, acidity, sorptive properties, and the like.

FINAL ASSUMPTIONS

With regard to the forest ecosystems, the sets of basic files are to meet the requirements of considering the forests as geographical and historical phenomena and in analyzing them as complex material systems. In analyzing the forest systems, the principle to be adopted is that these are functioning systems that have the ability of self-regulation. Consequently, they are systems that react in a definite manner to the changes occurring in the surrounding environment as well as to the presence and interference of man.

The system of influencing is defined by the knowledge of the system of functioning. The former system is to aim at improving the given biogeocoenosis. The improvement procedure applied, based on the knowledge of the attributes of the constituent elements of biogeocoenoses and their functioning, has the importance of an additional entopic factor, the direction and the intensity whereof may be consciously planned and controlled.

REFERENCES

Jäger, K. D. 1979. Aktuelle Fragen der Fachterminologie in der Periglazialforschung des nördlichen Mitteleuropa. Acta Univ. N. Cop., Geogr. (Toruń) 14(46):45-67.

Karte, J. 1979. Räumliche Abgrenzung und regionale Differenzierung des Periglaziärs. Bochumer Geogr. Arb. H.35. Paderborn. 211 pp.

Kopp, D. 1965. Die periglaziäre Deckzone (Geschiebedecksand) im nordostdeutschen Tiefland und ihre bodenkundliche Bedeutung. Ber. Geol. Ges. DDR, pp. 739-771.

_______. 1969. Die Waldstandorte des Tieflandes. Erg. d. Forstl. Standortserkundung in der DDR. Potsdam.

Kowaliński, S., R. Truszkowska, A. Kowalkowski, and J. Ostrowski. 1979. "BIGLEB" soil information system (general principles). Soil Sci. Ann. (Warsaw) 30(1):73-84. (In Polish.)

Kowalkowski, A. 1981. Kartographische Methode der Darstellung und Interpretierung der Struktur der Bodendecke als Standortskomponente. Mitt. Forstl. Bundesversuchsanst. Wien 140:57-64.

Kowalkowski, A., and J. Borzyszkowski. 1977. The role of periglacial and extraperiglacial perstructions in the formations of the soil profile in Central Europe. Folia Quaternaria (Krakow) 49:37-45.

Kurkin, N. A. 1973. Some methodological problems of studying the biogeocenoses and landscapes. In: Methodological problems of systemic studies, pp. 212-226. Warsaw. (In Polish).

Neef, E. 1967. Die theoretischen Grundlagen der Landschaftslehre. Gotha/Leipzig. 152 pp.

Part 2

NUTRIENT CYCLING AND SITE PRODUCTIVITY

9

NUTRIENT CYCLING IN WORLD FORESTS

D. W. Cole

INTRODUCTION

Our understanding of the rates and processes associated with nutrient cycling in the world's forests has increased dramatically in recent years. The International Biological Program (IBP) helped focus attention on nutrient cycling and the fundamental role it plays in forest ecosystems. Concerns related to the effects of acid rain and forest harvesting on future productivity of forest sites have resulted in expanded research in this area, as has interest in fertilizing forest sites to increase productivity and in recycling municipal sewage sludges and wastewater effluents in forest environments.

Such studies have added much to our knowledge of how forest ecosystems function and how species adapt to specific ecological niches. For example, the strategies of forest species in meeting annual uptake requirements for growth vary significantly from high elevations to low, from the northern latitudes to the tropics, and from deciduous to evergreen coniferous forests. Retention of foliage for up to twenty-five years by black spruce (*Picea mariana*) allows it to occupy sites where available nitrogen levels are exceptionally low. Similarly, remarkably high internal translocation of nitrogen from the foliage of larch (*Larix occidentalis*) before senescence allows it to compete with other coniferous species in marginal sites that again have a low supply of available nitrogen.

Here I discuss the processes regulating nutrient cycling in forest ecosystems, noting those that are common and critical to all forest types and those that are specific to unique situations. I also compare and discuss rates of cycling for various species, forest types, stand conditions, and regimes of the world. The relation between nutrient cycling and productivity of forest stands is discussed and comparisons are made between several forest regions.

PROCESSES REGULATING NUTRIENT CYCLING

Circulation of nutrients in a forest ecosystem follows specific pathways and moves at defined rates due to the inherent structure of the ecosystem and those processes that regulate flow rates. Different ecosystems and species within an ecosystem can assemble any number of strategies to meet nutrient uptake requirements and maintain a given rate of growth. In general, forest ecosystems have evolved mechanisms by which nutrients are conserved, and thus there is little nutrient loss from these ecosystems. The greater the deficiency of a nutrient, the more likely it is that this nutrient will be conserved and efficiently reused within the forest.

The general processes regulating nutrient flow within an ecosystem are (1) those associated with uptake by the higher plants, (2) translocation and reuse by these plants, (3) nutrient return to the soil surface, (4) mineralization and immobilization of returned nutrients by microorganisms, and (5) nutrient leaching through the soil profile. Of these, the most work reported has been in the area of nutrient return by litterfall. A simple comparison of such information between species or forest regions has little inherent value. The amount of nutrients that a tree returns to the soil each year through this process is dependent on other ecosystem processes and is poorly related to the nutrient needs of the tree, or its rate of primary production. For example, many coniferous species can vary the length of time that foliage is retained, depending on the abundance or deficiency of nutrients on the site. Deciduous species have a higher inherent rate of nutrient return than evergreen species. A comparison of these two forest types within the temperate region indicates that return of nutrients, except phosphorus, to the soil surface is nearly double in the deciduous stands (Table 1).

Table 1. Average litterfall and nutrient return (kg/ha/yr) for temperate deciduous and coniferous forests (adapted from Cole and Rapp 1980).

		Nutrient Return				
Forest Region	Litterfall	N	K	Ca	Mg	P
Temperate deciduous	5,400	61	42	68	11	4
Temperate coniferous	4,380	37	26	37	6	4

This difference is a result of the extended period of foliage retention by coniferous species. The retention time within coniferous species in the temperate region also varies, ranging from six to ten years for older Douglas-fir (*Pseudotsuga menziesii*) stands and from one to two years for shortleaf pine (*Pinus echinata*). Many species such as Douglas-fir also vary needle retention time depending on the nutrient status of the site and the age of the stand.

In addition, the return of nutrients to the soil surface does not closely follow litter return. Deciduous forests have, on average, 1.2 times more litter return and 1.6 to 1.8 times more nutrient return than coniferous forests. This is primarily because a greater percentage of returned biomass in a coniferous forest is not foliage but branches and other high-carbon, low-nutrient tissue. Also the pathway of return varies between elements. Data composited from the IBP studies suggest that nearly all returned nitrogen was associated with litterfall. However, the most potassium returned to the soil surface as stemflow and crown wash (Table 2).

Under some conditions the forest floor accumulates large quantities of nutrient elements primarily from litter deposition. Mean residence of nutrients in the forest floor layer is dependent on those factors that

Table 2. Pathways of nutrient return composited from all forest stands studied under the IBP (adapted from Cole and Rapp 1980).

Mechanism of Return	Nutrient Return				
	N	K	Ca	Mg	P
Litterfall	37	11	36	4	3
Stemflow and crown wash	6	17	11	3	1
Total return	43	28	47	7	4
% return in litterfall	83	41	71	60	85

regulate decomposition and on the specific element itself. For example, potassium, found largely in ionic form, has a shorter residence time in the forest floor than any other nutrient element. By contrast, nitrogen, which must be mineralized by microorganisms before it can be released, has a mean residence time about three times longer than potassium.

Since the process of decomposition largely governs residence time of nutrients in the forest floor, it follows that ecosystems at higher elevations and northern latitudes selectively accumulate far more nutrients within this compartment; Mediterranean and tropical ecosystems accumulate far less of a forest floor layer, and thus the residence time for these same elements is shorter. Table 3 compares forest floor

Table 3. Mean residence period in years for the forest floor and its nutrient constituents as found in major forest regions of the world (adapted from Cole and Rapp 1980; Turner and Singer 1976; Cole and Johnson 1979).

Region	Organic Matter	N	K	Ca	Mg	P
Boreal coniferous	350	230	94	150	455	324
Boreal deciduous	26	27	10	14	14	15
Subalpine coniferous	18	37	9	12	10	21
Temperate coniferous	17	18	2	6	13	15
Temperate deciduous	4	6	1	3	3	6
Mediterranean	3	4	<1	4	2	1
Tropical	0.7	0.6	0.2	0.3	--	0.6

residence times for the major nutrient elements and forest regions of the world. Since the nutrient status of the litterfall materials is higher, deciduous forests typically have far more rapid decomposition rates and thus shorter residence times for nutrient elements than coniferous forests located in the same region.

This rapid release of nutrient elements by the forest floor layers under deciduous forests results in high uptake rates. (A comparison of uptake rates by the major forest regions is made later in this article.) It should be recognized, however, that the processes leading to long nutrient residence times in the forest floor and uptake by vegetation minimize leaching potential and nutrient loss from the soil profile.

Besides uptake by the vegetative cover and slow decomposition of the forest floor, the principal means of conserving nutrients within the ecosystem are governed by those mechanisms that regulate leachate chemistry and leaching losses. Recent review articles by Vitousek (in prep.), Johnson and Cole (1980), and Cronan et al. (1978) discuss these mechanisms as they apply to forest ecosystems. These investigators hypothesized that nutrient loss through leaching processes is regulated by production and mobility of anions in the soil solution. By changing anion activity of the soil solution, cation leaching potential is changed by an equivalent amount. The principal anions regulating leaching losses vary between forest regions of the world. Johnson et al. (1977) compared leaching processes in tropical, temperate, subalpine, and northern latitude forest ecosystems. The dominant anion regulating leaching not only varied between sites but also by depth within a site. The major anions associated with leaching were typically bicarbonate, sulfate, and chloride. The nitrate ion becomes important in this leaching process when ammonium nitrogen concentrations are high, as in the red alder (*Alnus rubra*) ecosystems (Van Miegroet 1980), and under some conditions of forest disturbance (Vitousek et al. 1979; Cole et al. 1975; Likens et al. 1969). At higher elevations and at a northern latitude site in Alaska, Johnson et al. (1977) found that organic acids can also play an important role in this leaching process, especially within the surface soil horizons.

It is evident that the addition of mobile anions to a forest ecosystem (e.g., by fertilization, municipal sludge wastewater application, and acid precipitation) will theoretically accelerate rates of leaching if the added anions remain in the soil solution. However, any activity, natural or man-derived, that adds to anion loading of the soil solution can potentially result in increased losses from the ecosystem. At times anion levels are reduced by sorption on soil surfaces. This is especially true for phosphate and sulfate ions. The bicarbonate ion is automatically decreased with an increase in solution acidity and is not found to any meaningful extent below pH 4.5. The nitrate ion becomes abundant under conditions of high nitrification. Fortunately, such conditions are not common in forest ecosystems. However, they can occur as mentioned above, and only the process of uptake by the plant cover will significantly reduce this anion concentration in most forest conditions. The chloride anion is highly mobile and not readily taken up by the plant cover. Thus any process that adds this ion to the ecosystem, such as mineral weathering and ocean air, will result in a likely loss of cations through accelerated leaching. In all cases the extent of cation loss is stoichiometrically controlled by the anion composition of the soil solution.

The amount of leaching losses under most conditions is low and probably less than the ecosystem receives from the atmosphere, through biological fixation, and from mineral weathering processes. Table 4 summarizes such losses from those few sites where leaching rates have been monitored.

Table 4. Annual leaching losses (kg/ha) below the rooting depth under different ecosystem types (adapted from Cole and Rapp 1980).

Species	Site	N	K	Ca	Mg	P
Douglas-fir	Cedar River (U.S.)	0.6	1.0	4.5	-	.02
Spruce	Solling (W. Germany)	0.9	2.2	2.3	0.4	.06
Poplar	ORNL (U.S.)	3.5	8.9	44.5	-	.05
Beech	Solling (W. Germany)	6.0	2.9	12.7	3.7	.10
Oak, birch	Belgium	12.6	8.3	59.8	6.0	.20

RATES OF NUTRIENT CYCLING

Cycling rates vary between regions of the world and between species of a given region. This is largely because of inherent differences between species relative to nutrient requirements and cycling strategies. It is also a result of environmental differences between regions and forest types affecting nutrient availability, forest floor decomposition, and nutrient leaching losses. The effect of these processes, individually and collectively, on nutrient cycling rates within major ecosystem types is discussed next.

Effect of Stand Age

Rates of nutrient cycling, especially those associated with nutrient uptake and the nutrient amount associated with the growth increment for a given year (nutrient requirement) will vary with stand age (Table 5). With nutrients that are relatively mobile, such as nitrogen and

Table 5. Annual uptake and nutrient requirements of nitrogen, potassium, and calcium for Douglas-fir between the ages of 9 and 450 years (after Turner 1975; Grier 1974).

	N		K		Ca	
Age	Uptake	Requirement	Uptake	Requirement	Uptake	Requirement
9	4	6	4	4	5	2
22	33.7	42	26.3	31.0	34.4	13.3
30	32.1	33	29.5	27.7	55.7	14.4
42	32.8	36	27.4	26.1	40.9	12.9
73	32.5	35	21.4	24.5	43.2	14.1
95	37.3	29	25.5	25.9	51.4	11.9
450	23.7	35	21.2	26.7	53.3	17.9

potassium, uptake and nutrient requirement rates in Douglas-fir are nearly equal. Maximum uptake for these two elements is achieved approximately when the crowns of individual trees close and a uniform canopy develops. With calcium, however, a quite different pattern emerges with stand age. Annual uptake exceeds that required to produce one year's growth increment. This is caused by the immobility of this element together with its accumulation in older tissue. Since older-aged stands also retain foliage for longer periods, there is a greater opportunity for calcium to accumulate. Magnesium accumulates to an even greater extent that calcium in older tissue.

Effect of Nitrogen Fixation

Within a given forest region, striking differences in nutrient cycling can be found in species capable of symbiotic nitrogen fixation. In the temperate forest region, red alder (*Alnus rubra*) can fix from 50 to 100 kg/ha annually (Cole et al. 1978; Tarrant and Miller 1963; Newton et al. 1968).

Besides increasing the nitrogen status of the soil and the potential productivity of the ecosystem, fixation can also result in significant changes related to cycling. Typically, deciduous trees translocate about a third of the nitrogen in the foliage before litterfall. However, in red alder little if any nitrogen translocation occurs. Consequently, alder foliage remains on the tree longer in the fall than the foliage on other deciduous species growing in the same environment.

This increase in nitrogen status of a site due to fixation can also have potentially detrimental consequences. Nitrogen accumulating within the soil profile due to fixation results in a marked increase in the nitrification process and thus the soil nitrate concentration. The nitrate ion functions as a highly mobile source causing leaching losses of an equivalent number of soil cations. For an alder ecosystem in western Washington, nitrate is the dominant anion associated with leaching. Esssentially no nitrate is found in soil solutions of an adjacent Douglas-fir forest where fixation is not occurring (Van Miegroet 1980). Nitrate concentrations exceeding 10 ppm (drinking water standard established by the U.S. EPA) can at times be found in the soil solutions of alder ecosystems.

Differences Between Deciduous and Coniferous Ecosystems

Because the foliage of deciduous species is replaced each year (unlike that of most coniferous species), we would expect many aspects of mineral cycling between these major taxonomic groups to also be different. The IBP data on mineral cycling summarized by Cole and Rapp (1980) provide examples of such differences. It is apparent from this summary that the annual process of uptake, requirement, and return of five major nutrient elements is far more rapid in deciduous than in coniferous species (Table 6). For example, the nitrogen uptake rate of coniferous species is only 56% that of deciduous stands. The potassium, calcium, and magnesium uptake rates of coniferous species approximate half that of deciduous species. Only in the case of phosphorus is the difference relatively small (within 80%). Annual requirement and return of these elements also showed similar differences when compared between these two forest types.

The data in Table 6 also suggest that deciduous species translocate significantly more nitrogen from old to new tissue than do conifers for these IBP sites. Average nitrogen uptake for deciduous species is 70.5

Table 6. Comparison of deciduous and coniferous species relative to elemental uptake, requirement, and return (kg/ha/yr) (adapted from Cole and Rapp 1980).

Element	Uptake		Requirement		Return	
	Deciduous	Coniferous	Deciduous	Coniferous	Deciduous	Coniferous
Nitrogen	70	39	94	39	57	30
Potassium	48	25	46	22	40	20
Calcium	84	35	54	16	67	29
Magnesium	13	6	10	4	11	4
Phosphorus	6	5	7	4	4	4

± 21 kg/ha. However, 94 ± 19 kg/ha is needed to meet annual requirements for nitrogen in new tissue. Thus about a third of the annual requirement for nitrogen is met through translocation from old to new tissue. In the case of coniferous species, data suggest little if any translocation. Essentially identical quantities of nitrogen are taken up as are required to produce new tissue (39 ± 22 kg/ha/yr and 39 ± 21 kg/ha/yr). There is little apparent translocation of potassium for either deciduous or coniferous species. Calcium uptake exceeds the annual requirement for both deciduous and coniferous species. In the case of coniferous species, calcium uptake is about double the requirement. Much of the calcium taken up each year accumulates in older tissue and is not directly incorporated in new material. The pattern for magnesium is similar to that for calcium. Uptake greatly exceeds the annual requirement for both deciduous and coniferous species, suggesting that magnesium also accumulates in older tissues.

Rate of Nutrient Cycling in Forest Regions of the World

Accumulation and cycling of nutrient elements varies between forest regions of the world. This can be attributed to those processes discussed above. In northern latitudes, decomposition is slow, resulting in a large percentage accumulation of organic matter and nutrients in the forest floor layer. For example, about 70% of the organic matter above the soil surface for the boreal coniferous forest is contained within the forest floor layer. This percentage rapidly drops off to about 15% for temperate forests (Cole and Rapp 1980) and 3% for equatorial forests (Bartholomew et al. 1953). The distribution of nutrients between the vegetative cover and the forest floor follows approximately the same trend (Table 7). In the north regions about three-fourths of the nitrogen above the soil surface is contained in the forest floor layer. This percentage rapidly decreases to only 6% for an equatorial forest. Potassium follows the same trend with somewhat lower percentages because the leaching of this element from the forest floor layer does not depend on the decomposition process.

Annual uptake, requirement, and return of nitrogen by forest regions are summarized in Table 8. This comparison shows clearly that major differences in these aspects of aboveground cycling exist between

Table 7. Accumulation of nutrients (kg/ha) in the aboveground vegetation and forest floor by forest regions (adapted from Cole and Rapp 1980; Turner and Singer 1976; Bartholomew et al. 1953).

	Nitrogen			Potassium			Calcium		
Forest Region	Vegetation	Forest floor	%*	Vegetation	Forest floor	%*	Vegetation	Forest floor	%*
Boreal coniferous	117	617	84	44	109	71	258	360	58
Boreal deciduous	221	548	71	104	99	51	164	489	75
Subalpine coniferous	372	650	64	980	175	15	1,046	547	34
Temperate coniferous	479	681	59	340	70	17	680	206	23
Temperate deciduous	442	377	46	224	53	19	557	205	27
Mediterranean	745	125	14	626	10	2	3,753	301	9
Tropical	778	53	6	677	5	1	945	26	3

*Percentage contained in the forest floor.

forest regions. For example, annual uptake varies from a low of 2 kg/ha for spruce forests of boreal forests to about 160 kg/ha for equatorial forests (a value calculated from Bartholomew et al. 1953 and Cole and Johnson 1978). This high uptake value should not be surprising when we consider the high litter and nutrient return values generally reported for tropical areas (UNESCO 1978; Cole and Johnson 1978).

Except for tropical areas, an analysis of nitrogen uptake values would lead to the conclusion that trees are relatively modest in removing nitrogen from the soil. For the thirty-five IBP sites reported by Cole and Rapp (1980) the annual nitrogen uptake averaged 55 kg/ha.

Nitrogen requirement for annual growth closely follows uptake for coniferous forests and is somewhat higher than uptake for deciduous forests. This is primarily because of the generally high rates of translocation in deciduous trees before senescence of foliage. Although such information is not directly available for tropical forests, Nye (1961) speculated that tropical forests translocated less nutrients than those in temperate areas. This suggests that the nitrogen requirement of tropical sites approaches that of uptake.

The highest rate of nutrient return is found in tropical forests. From the five sites included in this summary (Table 8) (Cole and Johnson 1978), the average annual return of nitrogen was 114 kg/ha. Deciduous forests of the temperate region return only half this amount. The temperate coniferous forests return even less, 36 kg/ha/yr, because of needle retention.

Table 8. Average annual uptake, requirement, and return of nitrogen by forest region (kg/ha) (adapted from Cole and Rapp 1980; Cole and Johnson 1979; UNESCO 1978).

Forest Region	Uptake	Requirement	Return
Boreal coniferous	2	3	3
Boreal deciduous	25	56	20
Temperate coniferous	47	47	36
Temperate deciduous	75	98	61
Mediterranean	48	77	34
Tropical	160	--	114

RELATIONSHIP BETWEEN CYCLING AND FOREST PRODUCTIVITY

It is appropriate to ask the questions: What relationship, if any, exists between nutrient cycling and productivity of forest eocsystems? To what extent is the cycling noted above critical to the growth and structure of forest ecosystems? To what extent is it simply following other processes and thus has little significance to the productivity of these forest sites? It is difficult to obtain fully satisfactory and definitive answers to these questions. The relationship has been addressed in extensive reviews by Ovington (1962) and Rodin and Bazilevich (1967). Specific studies have compared elemental cycling, elemental distribution, and forest productivity for individual forest ecosystems with others that have been fertilized (Heilman and Gessel 1963; Fagerstrom and Lohm 1977; Madgwick et al. 1970). Since most forests of the world are nitrogen deficient, it is more likely that nitrogen uptake regulates production than vice versa. The literature supports this conclusion.

Assuming that uptake does indeed regulate forest production, Cole and Rapp (1980) calculated that 1 kg/ha/yr of nitrogen uptake by forest stands will, on average, annually produce 168 ± 49 kg/ha of aboveground biomass. In general, conifers seem to be more efficient than deciduous species. On average, 1 kg/ha/yr of nitrogen uptake will produce 143 ± 36 kg/ha/yr of deciduous biomass and 194 ± 48 kg/ha/yr of coniferous biomass. These differences between forest types are significant at the 99% confidence level and represent, in my opinion, inherent differences in efficiency of nitrogen utilization between these two major taxonomic groupings. Similar differences can also be seen between major forest regions of the world (Table 9). Again, some general relationships emerge from these data regarding the efficiency of nitrogen utilization by different forest types. Within the same region and with the same amount of nitrogen, coniferous forests are consistently more efficient than deciduous forests in producing biomass. The northern species seem to be more efficient in producing biomass for a given amount of nitrogen than are the more southerly species. This could be caused by a greater deficiency of available nitrogen in these boreal sites, resulting in a higher conversion ratio between production and nitrogen uptake and utilization.

Unfortunately, the data base for Table 9 is not uniform between regions; the vast majority of sites are located within temperate

Table 9. Aboveground biomass production per unit of nitrogen uptake (kg/ha/yr). (Number of sites in parentheses.)

Forest Region	Average Aboveground Production per kg of N Uptake
Boreal coniferous (3)	295
Boreal deciduous (1)	92
Temperate coniferous (13)	179
Temperate deciduous (14)	103
Mediterranean (1)	92
Tropical (7)	120

forests. The most questionable comparison is of values from the tropic sites. This vast forest region represents such an incredible array of species and site conditions that any generalization is speculative. In addition, few studies in the tropics have yielded enough productivity and cycling observations to calculate with confidence parameters such as uptake and efficiency of nutrient utilization. The data from UNESCO (1978), Gessel et al. (1980), and Cole and Johnson (1978) indicate that about 120 kg/ha of organic matter are produced for each 1 kg of nitrogen taken up in tropical sites. This value is reasonably close to that of deciduous forests of the temperate region.

The utilization of nitrogen by any forest ecosystem is exceptionally efficient when compared with other vegetative types (Table 10). This is

Table 10. Resource use efficiency of nitrogen by various ecosystems (Cole 1980).

Ecosystem Type	Aboveground Production per Unit of Nitrogen Uptake (kg/ha/yr)	Data Source
Forests	168 ± 49	Cole and Rapp 1980
Tundra	35	Stuart and Miller (submitted)
Mediterranean shrub	100 - 115	Gray and Schlesinger (in press)
Agricultural (corn)	25 - 60	Brawn et al. 1963
Bermuda grass	61 ± 30	Fisher and Caldwell 1959

true not only for tundra and Mediterranean shrub vegetation, but also for agricultural crops. There are obvious reasons for the high efficiency of nitrogen use by forests, including the following ones: (1) Forests, as perennial plant systems, can translocate and recycle nitrogen once it has been taken up. (2) Biomass of a forest has a higher C:N ratio than most annual plants. (3) Most forests are growing in nitrogen-deficient sites which favor high efficiency rates; better sites have already been converted to agricultural use.

Thus we can conclude that forests have selectively evolved in a nitrogen-deficient environment and are still the most efficient plant cover for such sites.

CONCLUSIONS

The information base for nutrient cycling in forest ecosystems of the world has reached a stage where meaningful comparisons between regions, forest types, and specific species can be made:

1. There are major differences in nutrient cycling between forest regions and species within regions. The rate of cycling, especially for nitrogen, is slower in northern latitudes than in Mediterranean and tropical regions. Within a given region, deciduous species cycle nutrients at higher rates than coniferous species. This generalization holds true for uptake, translocation, nutrient return, and forest floor decomposition.

2. In general, leaching losses from forest ecosystems are minor compared with nutrient storage and cycling rates within ecosystems. The basic rate of leaching is regulated by the production and mobility of soil solution anions. However, the specific anion involved in this leaching process varies between ecosystems, and by depth within an ecosystem.

3. Forest ecosystems are more efficient than other terrestrial plants in producing biomass for the amount of nitrogen taken up. In general, coniferous species are more efficient in this respect than are deciduous species. Forests of the boreal region are more efficient than those of lower latitudes.

REFERENCES

Bartholomew, W. V., J. Meyer, and H. Ladelout. 1953. Mineral nutrient immobilization under forest and grass fallow in the Yangambi (Belgian Congo) region. Publ. Inst. Nat. Agron. Congo Belge. Ser. Sci. No. 57. 27 pp.

Brawn, L. C., T. L. Nelson, and C. L. Crawford. 1963. Residual nitrogen from NH_4NO_3 fertilizer and from alfalfa plowed under. Agron. J. 55:231-235.

Cole, D. W. 1980. Nitrogen uptake and translocation by forest ecosystems. Ecol. Bull. (Stockholm). In press.

Cole, D. W., W. J. B. Crane, and C. C. Grier. 1975. The effect of forest management practices on water chemistry in a second-growth Douglas-fir ecosystem. In: B. Bernier and C. H. Winget, eds., Forest soils and forest land management, pp. 195-207. Les Presses de l'Université Laval, Quebec.

Cole, D. W., S. P. Gessel, and J. Turner. 1978. Comparative mineral cycling in red alder and Douglas-fir. In: D. G. Briggs, D. S. DeBell, and W. A. Atkinson, eds., Utilization and management of alder, pp. 327-336. USFS Pacific Northwest Forest and Range Experiment Station, Portland, Oregon.

Cole, D. W., and D. W. Johnson. 1978. Mineral cycling in tropical forests. In: C. T. Youngberg, ed., Forest soils and land use, pp. 341-356. Colorado State University Press, Fort Collins.

Cole, D. W., and M. Rapp. 1980. Elemental cycling in forest ecosystems. In: D. E. Reichle, ed., Dynamic properties of forest ecosystems, pp. 341-409. Cambridge University Press, New York.

Cronan, C. S., W. A. Reiners, R. C. Reynolds, Jr., and G. E. Long. 1978. Forest floor leaching: Contributions from mineral, organic and carbonic acids in New Hampshire subalpine forests. Science 200:309-311.

Fagerstrom, T., and U. Lohm. 1977. Growth in Scots pine (*Pinus silvestris* L.). Oecologia (Berlin) 26:305-315.

Fisher, F. L., and A. G. Caldwell. 1959. The effect of continued use of heavy rates of fertilization on forage production and quality of coastal Bermuda grass. Agron. J. 51:99-102.

Gessel, S. P., D. W. Cole, D. Johnson, and J. Turner. 1980. The nutrient cycles of two Costa Rican forests. Progress in Ecology 3:23-44.

Gray, J. T., and W. T. Schlesinger. In press. Nutrient cycling in Mediterranean type ecosystems. In: P. C. Miller, ed., Resource utilization in Mediterranean type ecosystems. Springer-Verlag.

Grier, C. C., D. W. Cole, C. T. Dyrness, and R. L. Fredriksen. 1974. Nutrient cycling in 37- and 450-year-old Douglas-fir ecosystems. In: Integrated research in the coniferous forest biome, pp. 21-34. Coniferous Forest Biome Bull. 5. University of Washington, Seattle.

Heilman, P. E., and S. P. Gessel. 1963. Nitrogen requirements and the biological cycling of nitrogen in Douglas-fir stands in relationship to the effects of nitrogen fertilization. Plant and Soil 18:386-402.

Johnson, D. W., and D. W. Cole. 1980. Anion mobility in soils: Relevance to nutrient transport from forest ecosystems. Environ. Int. 3:79-90.

Johnson, D. W., D. W. Cole, S. P. Gessel, M. J. Singer, and R. V. Minden. 1977. Carbonic acid leaching in a tropical, temperate, subalpine and northern forest soil. Arctic and Alpine Res. 9:329-343.

Likens, G. E., F. H. Bormann, and N. M. Johnson. 1969. Nitrification: Importance to nutrient losses from a cut-over forested ecosystem. Science 163:1205-1206.

Madgwick, H. A. I., E. H. White, G. K. Xydias, and A. L. Leaf. 1970. Biomass of *Pinus resinosa* in relation to potassium nutrition. For. Sci. 16(2):154-159.

Newton, M., B. A. el-Hassan, and J. Zavitkovski. 1968. Role of red alder in western Oregon forest succession. In: J. M. Trappe, J. F. Franklin, R. F. Tarrant, and G. M. Hansen, eds., Biology of alder, pp. 73-84. Pacific Northwest Forest and Range Experiment Station, Portland, Oregon.

Nye, P. H. 1961. Organic matter and nutrient cycles under moist tropical forest. Plant and Soil 13:333-346.

Ovington, J. D. 1962. Quantitative ecology and the woodland ecosystem concept. In: Advances in ecological research, 1:103-192.

Rodin, L. E., and N. I. Bazilevich. 1967. Production and mineral cycling in terrestrial vegetation. Oliver and Boyd Ltd., London. 288 pp.

Stuart, L., and P. C. Miller. Submitted. Seasonal nutrient and biomass accumulation of the vegetation on undisturbed sites and in a vehicle track in _Eriophorum vaginatum_ tuska tundra.

Tarrant, R. F., and R. E. Miller. 1963. Accumulation of organic matter and soil nitrogen beneath a plantation of red alder and Douglas fir. Soil Sci. Soc. Am. Proc. 27:231-234.

Turner, J. 1975. Nutrient cycling in a Douglas-fir ecosystem with respect to age and nutrient status. Ph.D. thesis, University of Washington, Seattle.

Turner, J., and M. J. Singer. 1976. Nutrient distribution and cycling in a subalpine coniferous forest ecosystem. J. Appl. Ecol. 13:295-301.

UNESCO. 1978. Tropical forest ecosystems. UNESCO/UNEP/FAO, Paris.

Van Miegroet, H. 1980. Effects of urea fertilization and sawdust amendment on microbial activity and nutrient leaching to two soil systems with different nitrogen status. M.S. thesis, University of Washington, Seattle. 148 pp.

Vitousek, P. M., J. R. Gosz, C. C. Grier, J. M. Melillo, W. A. Reiners, and R. L. Todd. 1979. Nitrate losses from disturbed ecosystems. Science 204(4):469-474.

10

NUTRITIONAL COSTS OF SHORTENED ROTATIONS IN PLANTATION FORESTRY

R. J. Raison and W. J. B. Crane

INTRODUCTION

One of the most challenging questions facing forest research is whether we can sustain (or improve) the productivity and integrity of today's forests for tomorrow. There is ample evidence from past civilizations testifying to the importance of maintaining stable productive systems. As Rene Dubois (1973) has written: "Diseases, warfare and civil strife have certainly played important roles in the collapse of ancient civilizations; but the primary cause was probably the damage caused to the quality of the soil and to water supplies by poor ecological practices."

The probable pattern of forest development for the next half century is becoming increasingly clear. For the developed forested countries, the readily accessible old-growth forests are largely or completely cutover, and the intensity with which the regrowth forests are being managed is rapidly increasing. Intensified management includes the use of selected fast-growing species (often grown in monoculture), greater site preparation and manipulation (including cultivation, weed and pest control, fertilization, and in some cases irrigation), more frequent thinning, shortening of rotations, and more complete harvesting of biomass. Some countries (such as Sweden) are now approaching optimal economic productivity of their native forests, and in the face of continually increasing demand for wood, have turned to more intensive (plantation) forestry, often in countries outside their own boundaries (such as in Brazil).

If intensity of management is considered as a scale in which productivity for a specific climate and soil stretches from a "native" unmanaged productivity to a biologically possible maximum, then the change from extensive management of native forest to intensive plantation management is considerable. The productivity of even-aged plantations that are clearcut prior to maturity can be considerably greater than that of the native forests that preceded them. This is particularly true in Australia. However, plantations of introduced species may be less well adapted in the long term to the soils and climatic conditions of the site. The increased growth rates and removal of biomass will increase the demand on the soil for nutrients, and more disturbance of the site may markedly increase the potential for associated losses of organic matter and nutrients from the forest ecosystem. A fundamental long-term question of forest management must be: what are the most efficient strategies to ensure a sustained supply of water and nutrients to the growing crop? The answer will require an understanding and quantification of various options of management.

The consequences for tree nutrition of shortening rotations can be broadly divided into two categories: (1) direct effects on the rate of nutrient drain from the site, and (2) indirect effects associated with harvesting, site preparation, and development of the new stand.

A detailed review of the effects of shortening rotations in different forest types is not attempted here (much relevant information is contained in Leaf 1979). Rather the significant facts that have emerged to date are summarized, and some current gaps in knowledge are highlighted.

MANAGEMENT OPTIONS DIRECTLY AFFECTING REMOVAL OF NUTRIENTS IN BIOMASS

The amount of and rate at which nutrients are directly exported from a forest in biomass depend on combinations of three factors: (1) the species group, (2) the length of rotation, and (3) the degree to which biomass is utilized.

Increasing demand for forest products and new technologies has stimulated adoption of both increased levels of biomass utilization and shortening of rotations. Sometimes both strategies are applied together, leading to what has been termed "fiber farming" (Young 1975). Increased utilization of biomass (such as whole-tree harvesting in which all the aboveground biomass is removed, or complete-tree harvesting in which the stumps and roots are also included) can increase the amount of specific nutrients removed from the forest by up to fivefold (Malkonen 1976; Kimmins 1977). The percentage increase in nutrient removal with whole-tree logging compared with conventional logging is greatest when harvesting young stands where relatively more of the nutrients in aboveground biomass are stored in the components of the canopy (Kimmins 1977). The shortening of rotations also increases the rate at which nutrients are removed in harvested biomass (Crane and Raison 1981), especially where the degree of utilization is high (Boyle and Ek 1972). In each of these cases (when the level of utilization is increased, and where rotational length is decreased), the nutritional efficiency (i.e., the mass of biomass exported per mass of nutrient harvested) is reduced (and the nutritional cost increased), compared with a more traditional management (harvesting of logs only, possibly by selection logging, and management on longer rotations).

There can be a marked interaction between the type of tree species and the length of rotation that will affect the rate of nutrient export and nutritional efficiency. A study of the distribution of biomass and phosphorus (P) in the boles (wood + bark) in two commercially significant species, *Eucalyptus delegatensis* and *Pinus radiata*, in Australia led to the following conclusions (Crane and Raison 1981):

1. A markedly lower concentration of P was found in heartwood than in sapwood for both species. This was shown especially in *E. delegatensis*, where the ratio of P in the sapwood to that in the heartwood was 33:1, compared with a ratio of 9:1 in *P. radiata*.
2. Significantly more sapwood formation was found in *P. radiata*. Heartwood formation did not commence in *P. radiata* until age 17, compared with *E. delegatensis*, where heartwood commenced to form at age 7.
3. The combination of faster growth rates and a higher average concentration of P in the boles--a consequence of (1) and (2) above--resulted in greater rates of P removal when harvesting

P. radiata compared with E. delegatensis. These rates are shown in Table 1.

4. Independent of growth rate, the shortening of rotations increased the removal of P per unit of wood harvested. Table 1 shows that the amount of P removed per unit of wood harvested increased by 70% when the rotational age of E. delegatensis was reduced from 57 to 18 years. Removal of early thinnings (a form of shortened rotation management) results in a significant loss of nutrients from the P. radiata forest.
5. Theoretically, E. delegatensis will only become significantly more nutritionally efficient than P. radiata when grown on rotations of greater than seven years. The comparative efficiency of E. delegatensis increased progressively with age.

This latter observation has led us to conclude that for rotations of about seven years or less which are currently used with Eucalyptus plantations in Brazil, there may be no nutritional advantage (in reduced export of P in biomass) of eucalypts over other species as suggested by Hillis and Brown (1978). Eucalypt plantations exhibiting high rates of growth (40 to 80 m^3/ha/yr) outside of Australia, and managed on short (< 10 yr) rotations, will place heavy demands on soil reserves of nutrients (e.g., ∿ 5 kg P/ha/yr stored in wood), which are similar to

Table 1. Quantities and rates of phosphorus exported from the forest when harvesting E. delegatensis and P. radiata on short and long rotations (after Crane and Raison 1981).

Parameter	Eucalypt (E. delegatensis)		Pine (P. radiata)	
Tree rotation (yr)	18	57	18	40*
Proportion of sapwood in stemwood (%)	52	28	100	90
P exported (kg P/ha):				
in stemwood	9	17	28	56
in bark	4	8	18	24
in bole	13	25	46	80
Rates of P harvested in boles:				
as per wood (g P/t wood)	97	51	258	169
as per time (kg P/ha/yr)	0.73	0.44	2.53	1.97
Proportions of the rates of P harvested in boles compared with longer rotation:				
as per wood	1.7	1	1.5	1
as per time	1.6	1	1.3	1

*Includes four commercial thinnings at ages 16, 22, 28, and 34 prior to clearcutting at age 40.

those of annual agricultural crops (e.g., 7+ kg P/ha/yr for corn; Hanway and Olson 1980). The harvesting of foliage in addition to wood will further increase the removal of nutrients, and it is very likely that regular fertilization (at rates similar to those used for agricultural crops) will be required if fertility of the soil and productivity of the forest are to be sustained.

INDIRECT EFFECTS OF SHORTENED ROTATIONS ON NUTRIENT LOSSES

Shortened rotations clearly increase the frequency of disturbance to the site that results from harvesting and site preparation. The proportion of time the site is not fully occupied by a closed-canopy forest (which offers some protection to the soil) will also increase. Site disturbance will usually lead to either direct loss or increased potential for loss of nutrients from the site. The relative magnitude of nutrient export from a site in biomass compared with that associated with site disturbance is highly variable. However, the latter can sometimes be greater than the former. The effects of disturbances on losses of nutrients via leaching, erosion, and transfer to the atmosphere, and on undesirable redistributions of nutrients and compaction of soils are discussed below.

Soil Erosion

The rate of soil erosion on a forested site is difficult to measure, and assessment of the consequences of soil loss for subsequent productivity is more difficult (McColl and Grigal 1979). It is clear that accelerated erosion is undesirable because the most nutrient-rich and best-structured soils are usually removed. Redistribution of soil downslope from ridges to near-stream locations (e.g., into streamside reserves or "buffer" strips) is undesirable from the standpoint of tree growth even though the soil may not be exported from the catchment. Removal of 1 cm of nutrient-rich surface soil results in the export of about 24 kg P and 240 kg N/ha (assuming soil contains 0.3% N and 0.03% P, bulk density = 0.8 g/cm^3). Even modest rates of soil erosion during a forest rotation may result in losses of nutrients of the same order of magnitude as occur in harvested biomass. Long-term and well-designed field studies are needed to evaluate the effect of erosion on subsequent growth rates of trees.

Leaching

Harvesting, which invariably results in some degree of disturbance of the soil and litter layer, often leads to increased leaching of nutrient ions. Factors such as the removal of vegetation, which is a sink for both water and nutrients, and increased exposure and associated soil respiration rates that can enhance the production of mobile anions (especially bicarbonate and nitrate) are involved. The magnitude of leaching losses following clearcutting and site preparation is highly variable, but losses are usually greatest for N (as NO_3^-) and cations (K, Ca, Mg). Commercial clearcutting can result in losses of calcium and nitrogen amounting to 30% to 50% or more of those removed in conventional harvesting of biomass (Likens et al. 1978). Normally, however, losses are of a lesser magnitude (Wells and Jorgensen 1979). Leaching rates usually return to preharvest levels within three to five years of logging (McColl and Grigal 1979). The effects of shortened rotations on leaching rates have not been reported, but average long-term rates would be expected to increase because of more frequent disturbance of the site and an increase in the proportion of time when the site is not fully occupied by actively growing vegetation. Increased leaching further aggrevates the increased drain on site nutrients imposed by frequent harvesting of nutrient-rich biomass under shorter-rotation management.

Atmospheric Transfer

Burning of slash between rotations leads to some direct loss of nutrients contained in the fuel to the atmosphere (e.g., Harwood and Jackson 1975; Kimmins and Feller 1976). In addition, there can be other losses, especially of volatile elements, from duff layers and organically enriched surface soil during the fire, as well as subsequent loss of fire-mobilized elements from the soil (e.g., Gagnon 1965; Raison 1979). Losses of elements, especially nitrogen, phosphorus, and sulfur, resulting from burning may be equal to or greater than those removed in harvested biomass (e.g., Raison 1980, 1981). Although more complete utilization of biomass directly increases export of nutrients from a site, such procedures may eliminate the need for slash burning and hence prevent large losses of elements (especially nitrogen) from duff and surface soils.

Increased soil respiration following exposure of the soil at the time of harvesting can also lead to the loss of soil organic matter. This is in fact a transfer to the atmosphere independent of firing. Serious reduction of up to 60% in the organic matter content of soils following cropping with *P. radiata* plantations in Australia has been shown by Hamilton (1965), together with a widening of C/N ratios. Such a loss of organic matter is likely to have serious nutritional consequences because soil organic matter is so intrinsically associated with most physical, chemical, and biological aspects of soil fertility.

Nutrient Redistribution and Soil Compaction

Windrowing, which is often used during site preparation, can lead to scalping and removal of surface soil with detrimental effects on subsequent nutrient supply and tree growth (Ballard 1978). The degree and seriousness of soil disturbance and compaction caused by traffic of machines during harvesting and site preparation is highly variable (see Greacen and Sands 1980), but shortened rotations increase the potential for loss of site production as a result of these factors. Loss in production may result both from a reduction in the area on which trees can be established and from reduced growth rates of individual trees.

STRATEGIES FOR MANAGEMENT AND RESEARCH

Although depletion of site nutrients accompanies many intensive forestry practices, there are only rare documentations of a decline in the productivity of successive rotations (e.g., Ballard 1979). This does not mean that nutrient depletion may be unimportant, but reflects more the lack of experimentation, the difficulties in comparing growth rates of rotations growing decades apart, and the concentration of many shorter-term studies on "young" soils in environments with high accessions of nutrients in precipitation. Longer-term effects on poorer sites (where both soil reserves and weathering inputs are less) and in unpolluted environments may lead to different conclusions. Many plantation areas are, of course, only just commencing the second or third rotation under intensive silviculture, so the cumulative effects of cropping on productivity have had little opportunity to be expressed.

Natural nutrient inputs into forests are normally insufficient to balance losses associated with intensive short-rotation harvesting (Kimmins 1977; Shoulders and Wittwer 1979; Wells and Jorgensen 1979), or to supply nutrients at a sufficient rate during periods of peak demand (which usually occur prior to canopy closure). Fertilizer additions will be needed to replace losses where deficiencies already exist, or where these are induced. The effectiveness and efficiency with which fertilizers can replace nutrients are poorly understood; but the recovery (or efficiency) of some applied elements (such as nitrogen) in biomass is often < 30% (Ballard 1979). The role of topsoil and organic residues in providing exchange sites, microbial substrates, buffering of nutrient supply, and storage of moisture cannot be provided directly by mineral fertilizers. Maintenance of nitrogen supply appears to be more difficult than that for other elements in intensively managed forests.

Fertilization aften leads to increased concentrations of the added elements in tree biomass (e.g., Will 1965; Schmitt et al. 1979; Cromer et al. 1980), so that export of nutrient per unit of biomass harvested will be greater where fertilizer is a component of a management regime. This effect may be most extreme in very short-rotation (< 10 yr) plantation systems.

Maintenance of soil organic matter is critical for maintenance of long-term soil fertility. The only alternative is the adoption of a costly (and often environmentally undesirable) nutritional system based mainly on heavy application of inorganic fertilizers, and which in extreme cases may approach sand culture. The difficulties of such systems derive from the problems of matching nutrient availability to tree uptake. Organic matter buffers against nutrient loss. Replacement of organic matter is both difficult and expensive. Use of fallow periods during which manure crops are grown may be needed in extreme cases, or intercropping of legumes during the early years of the new stand could be valuable especially in very sandy soils, which are inherently low in organic matter. Periods of lost production (fallow), and costs of establishing and maintaining legumes, can all be classified as nutritional costs, and may be needed for successful short-rotation forestry.

Sufficient research has now been carried out to enable reasonable prediction of the amounts of nutrients exported in biomass for a range of environments and forest management options. Efforts are still needed to quantify other losses of nutrients associated with short-rotation forestry management (such as through erosion, leaching, and gaseous tranfers) and in particular to assess their significance for long-term productivity of sites. The latter requires much more study of soil processes which control the rate of nutrient supply and a better understanding of the changing demands and sources of elements occurring with stand development. The above information will provide an improved basis for planning management regimes that minimize losses of nutrients per unit of biomass produced, and that make most efficient use of both natural and artificially applied nutrients. The impact of harvesting on soil compaction and subsequent root growth is also an area that demands high research priority.

The following strategies would appear important in order to maintain efficient long-term nutrient supply in short-rotation forestry programs: (1) leaving nutrient rich biomass (fine branches and foliage) on the site, and not harvesting root systems on most sites; (2) use of conservative site preparation procedures that minimize disturbance and loss of nutrients and organic matter from slash, litter layers, and surface soil (e.g., retention of slash where possible, avoidance of redistribution of topsoil during windrowing, planning of vehicular traffic, and use of procedures that minimize the overland flow of water); (3) efficient use of fertilizers based on an understanding of tree requirements and fate of applied nutrients; (4) the possible use of legumes to assist in the maintenance of soil organic matter and nitrogen economy; and (5) use of tree stock selected for its low nutrient demand (e.g., Forrest and Ovington 1971; White and Harvey 1979).

The efficient application of all these strategies depends on an improved understanding of the factors (not just nutritional) controlling the productivity of forest stands under a range of environmental conditions. The factors involved will be quite variable, as will the strategies needed to maintain adequate long-term rates of nutrient supply under short-rotation forest cropping. To date such studies have mostly concentrated on the more inherently fertile soils. Increased research effort is needed on relatively infertile soils (e.g., in Australia) where the effects of intensive harvesting on subsequent productivity are likely to be more marked, and the maintenance of stable highly productive plantations a continuing challenge for forest managers.

REFERENCES

Ballard, R. 1978. The effect of slash and soil removal on the productivity of second rotation radiata pine on a pumice soil. N.Z. J. For. Res. 8:248-258.

__________. 1979. Use of fertilizers to maintain productivity of intensively managed forest plantations. In: Impact of intensive harvesting on forest nutrient cycling, pp. 321-342. Proc. Symp. State University of New York, College of Environmental Science and Forestry, Syracuse. 421 pp.

Boyle, J. R., and A. R. Ek. 1972. An evaluation of some effects of bole and branch pulpwood harvesting on site macronutrients. Can. J. For. Res. 2(4):407-412.

Crane, W. J. B., and R. J. Raison. 1981. Removal of phosphorus in logs when harvesting _Eucalyptus delegatensis_ and _Pinus radiata_ forests on short and long rotations. Aust. For. (in press).

Cromer, R. N., E. Williams, and D. Tompkins. 1980. Biomass and nutrient uptake in fertilized E. globulus. Paper presented to IUFRO Symposium and Workshop on Genetic Improvement and Productivity of Fast Growing Trees, Brazil, August 1980.

Dubois, R. J. 1973. Humanizing the earth. Science 179:769-772.

Forrest, W. G., and J. D. Ovington. 1971. Variation in dry weight and mineral nutrient content of Pinus radiata progeny. Sil. Genet. 20:174-179.

Gagnon, J. D. 1965. Nitrogen deficiency in the York River burn, Gaspe, Quebec. Plant and Soil 23:49-59.

Greacen, E. L., and R. Sands. 1980. Compaction of forest soils: A review. Aust. J. Soil Res. 18:163-189.

Hamilton, C. D. 1965. Changes in the soil under Pinus radiata. Aust. For. 29:275-289.

Hanway, J. J., and R. A. Olsen. 1980. Phosphate nutrition of corn, sorghum, soybeans and small grains. In: The role of phosphorus in agriculture, pp. 681-692. Proc. Symp. Am. Soc. Agron., Alabama.

Harwood, C. E., and W. D. Jackson. 1975. Atmospheric losses of four plant nutrients during a forest fire. Aust. For. 38:92-99.

Hillis, W. E., and A. G. Brown. 1978. The need for improved wood production from eucalypts. In: W. E. Hillis and A. G. Brown, eds., Eucalypts for wood production, p. 343. CSIRO, Australia.

Kimmins, J. P. 1977. Evaluation of the consequences for future tree productivity of the loss of nutrients in whole-tree harvesting. For. Ecol. Manage. 1:169-183.

Kimmins, J. P., and M. C. Feller. 1976. Effect of clearcutting and broadcast slashburning on nutrient budgets, streamwater chemistry and productivity in western Canada. XVI IUFRO World Congress, Norway, Division 1, pp. 186-198.

Likens, G. E., F. H. Bormann, R. S. Pierce, and W. A. Reiners. 1978. Recovery of a deforested ecosystem. Science 199:492-496.

Malkonen, E. 1976. Effect of whole-tree harvesting on soil fertility. Silva Fenn. 10:157-164.

McColl, J. G., and D. F. Grigal. 1979. Nutrient losses in leaching and erosion by intensive forest harvesting. In: Impact of intensive harvesting on forest nutrient cycling, pp. 231-248.

Raison, R. J. 1979. Modification of the soil environment by vegetation fires, with particular reference to nitrogen transformations: A review. Plant and Soil 51:73-108.

__________. 1980. Possible forest site deterioration associated with slash-burning. Search 11:68-72.

__________. 1981. More on the effects of intense fires on the long-term productivity of forest sites: Reply to comments. Search 12:10-14.

Schmitt, M. D. C., M. M. Czapowskyj, L. O. Safford, and A. L. Leaf. 1979. Biomass and elemental content of fertilized and unfertilized _Betula papyrifera_ Marsh. and _Populus gradidentata_ Michx. In: Impact of intensive harvesting on forest nutrient cycling, p. 416.

Shoulders, E., and R. F. Wittwer. 1979. Fertilizing for high fiber yields in intensively managed plantations. In: Impact of intensive harvesting on forest nutrient cycling, pp. 343-359.

Wells, C. G., and J. R. Jorgensen. 1979. Effect of intensive harvesting on nutrient supply and sustained productivity. In: Impact of intensive harvesting on forest nutrient cycling, pp. 212-230.

White, E. H., and A. E. Harvey. 1979. Modification of intensive management practices to protect forest nutrient cycles. In: Impact of intensive harvesting on forest nutrient cycling, pp. 264-278.

Will, G. M. 1965. Increased phosphorus uptake by radiata pine in Riverhead forest following superphosphate applications. N.Z. J. For. 10:33-42.

Young, H. E. 1973. Biomass nutrient elements harvesting and chipping in the complete tree concept. 10th Res. Conf. API-Tappi College Relations Group, Point Clear, Alabama, November 8-10.

11

MAINTENANCE OF PRODUCTIVITY OF RADIATA PINE MONOCULTURES ON SANDY SOILS IN SOUTHEAST AUSTRALIA

P. W. Farrell, D. W. Flinn, R. O. Squire, and F. G. Craig

INTRODUCTION

The productivity of forests is determined by many biotic and abiotic factors and by the interaction of these factors through time (Switzer 1978). Productivity can be expressed as total biomass or, in forests managed for wood production, as volume growth over a specified rotation length. In plantation forestry the major objective is to maximize wood production, and the forest manager attempts to achieve this by manipulating various biotic and abiotic factors. But successful manipulation is dependent on an understanding of the role of these factors in the ecosystem and the effects of various management practices on them.

Maintenance of productivity in radiata pine (*Pinus radiata* D. Don) monocultures has been of serious concern to forest managers, particularly where plantations have been established on infertile, sandy soils low in organic matter, such as those in southeast South Australia and adjacent areas in southwest Victoria (Flinn et al. 1980). This concern arose from reports by Keeves (1966) and Bednall (1968) of a decline in productivity between first-rotation (1R) and second-rotation (2R) stands of radiata pine in this region. Consequently a major research program was initiated (Keeves 1966), with traditional silvicultural practices and philosophies receiving close scrutiny.

The Forests Commission Victoria added further impetus to this program in 1972 with the initiation of a comprehensive research project at Rennick in southwest Victoria. This project had three main objectives: (1) definition of the magnitude of productivity changes between successive rotations, (2) identification of the causes of any observed productivity changes, and (3) development of silvicultural practices that would maintain or increase productivity over successive rotations.

Results from this project have been reported by Flinn (1978), Flinn et al. (1979, 1980), Hopmans et al. (1979, 1980), Squire et al. (1979), and Leitch and Farrell (1980). This paper reviews these results and presents some current data from the experiment reported by Squire et al. (1979) and a hitherto unpublished study by the senior author. The implications of the various findings in relation to plantation establishment practices are discussed.

THE STUDY AREA

Rennick Plantation is located in southwest Victoria (latitude 37° 49', longitude 140°59'), within 25 km of the southern coastline and 18 km of Mount Gambier, South Australia. Over 100,000 hectares of radiata pine plantations have been established in this general region, mostly on sandy soils of aeolian origin. These soils have been described in detail by Stephens et al. (1941). Topography is flat, with the only relief provided by the remnants of dune crests and troughs.

Research undertaken by the Forests Commission Victoria has been confined to stands of radiata pine growing on Caroline, Kromelite, and Wandilo sands (Stephens et al. 1941). These sands are relatively infertile compared with many forest soils, with low organic matter levels in the surface horizons also being an important feature (Flinn et al. 1980). In a study comparing soil chemical properties under native eucalypt forest growing on Caroline sand and mature radiata pine growing on Caroline and Kromelite sands at Rennick, Hopmans et al. (1979) found that the surface horizon (0-25 cm) contained around 1.4% to 1.7% organic matter, 0.03% to 0.04% total N, 37 to 50 ppm total P, and 1.2 to 1.5 ppm water soluble P.

The area has a mean annual rainfall of around 800 mm, with pronounced summer minimum and winter maximum. Soil moisture availability, particularly during summer and autumn, and nutrient availability are major factors regulating the growth of radiata pine on these sandy soils (Ruiter 1979).

QUANTIFYING PRODUCTIVITY CHANGES BETWEEN SUCCESSIVE ROTATIONS

Comparisons of productivity between successive rotations are usually made using either paired plots on different sites (e.g., Whyte 1973; Evans 1975) or plots on precisely the same site through time (e.g., Keeves 1966); each approach presents unique problems. The paired-plot approach suffers from the difficulty of precisely matching edaphic and microclimatic factors between sites. On the other hand, where comparisons are made through time on the same site, controlling genotype and exactly replicating site preparation, planting and tending techniques between successive rotations can be difficult, and the results are subject to climatic variation.

In the study reported by Squire et al. (1979) an attempt was made to combine the best features of both approaches. The study encompassed a range of site qualities and included comparisons of 1R and 2R growth on precisely the <u>same</u> sites and an examination of growth on <u>matched</u> 1R and 2R sites in relation to edaphic, physiological, and meteorological factors. A detailed account of the methods used to establish the study is given by Squire et al. (1979). The essential features are given below.

For the first part of this study, plots were established in both low-site quality (SQ V based on Lewis et al. 1976) and high-site quality (SQ II) 1R stands of 25- to 28-year-old radiata pine in the Rennick Plantation. The stands had been established on native forest sites that were hand-cleared and the debris broadcast burned prior to planting. Total underbark volume and height growth as functions of age were determined for sample trees by stem analysis before the stands were clearcut and the 2R crop planted. Logging residue from the first crop was left unburned and distributed uniformly over the plots. The nonburning of residue before establishing the 2R crop was the only major deviation in site preparation between the two rotations. In all other aspects every attempt was made to establish the 2R crop in exactly the same manner as the 1R crop. Seeds were collected from the 1R plots before clearcutting, and seedlings were raised under cultural practices used in the late 1940s. Past practice was also followed during planting. Climatic differences between rotations are being measured.

In the second part of the study, plots were established in native forest sites adjacent to the 2R plots described above. These 1R sites were matched with the 2R sites on the basis of species composition and height of native vegetation, topographic position, and soil profile characteristics. To simulate past practice the native vegetation was hand-cleared, broadcast burned, and the plots pit-planted using seedlings raised as described earlier. Both parts of the study were established in 1975.

Comparisons of mean cumulative height and basal area increments on matched 1R and 2R plots at approximately yearly intervals since planting (Table 1), based on an approximate 3% subsample as described by Squire

Table 1. Mean cumulative height and basal area (at 10 cm above ground level) increments for matched 1R and 2R plots.

	Years					
Treatment	0.1	1.1	2.2	3.1	4.1	5.0
			Cumulative Height Increment (m)			
1R - Low SQ	0.00	0.10	0.40	0.64	1.33	2.43
1R - High SQ	0.00	0.10	0.43	0.73	1.54	2.77
2R - Low SQ	0.00	0.25	0.96	1.61	3.20	5.15
2R - High SQ	0.00	0.27	0.98	1.46	2.68	4.57
LSD ($P < 0.05$)	--	0.06	0.23	0.38	0.62	0.90
2R/1R CAI		2.6	2.3	2.1	1.9	1.7
			Cumulative Basal Area Increment (cm^2)			
1R - Low SQ	0.0	0.2	1.1	3.8	14.0	36.9
1R - High SQ	0.0	0.2	1.2	4.8	18.4	46.3
2R - Low SQ	0.0	0.8	8.6	26.0	67.1	122.7
2R - High SQ	0.0	0.9	7.4	21.0	53.8	106.1
LSD ($P < 0.05$)	--	0.4	3.6	9.5	21.3	34.3
2R/1R CAI		4.4	7.7	4.9	3.1	2.1

et al. (1979), indicate that growth on 2R sites is significantly better ($P < 0.05$). These differences were established within the first year of planting and will continue to increase while the ratio of current annual increment (CAI) on 2R to 1R plots remains above 1.0.

The first statistical examination of growth data between 1R and 2R crops on the same sites was made at age five years. Preliminary results are consistent with the analysis of growth on matched 1R and 2R plots. Comparisons of total underbark volume (Squire 1982) indicated that 2R growth had not declined, and in fact had increased on the low quality sites. Total underbark volume on the low quality sites in the first rotation was around 6 cubic meters per hectare compared with around 30 cubic meters per hectare in the second rotation. Volume growth on the matched 1R plots was around 5 cubic meters per hectare, which compares closely with that measured in the first rotation on the 2R sites.

SOME FACTORS IMPORTANT TO THE MAINTENANCE OF SITE PRODUCTIVITY

Flinn et al. (1980) concluded that the productivity of radiata pine on the infertile sandy soils typical of the general region of this study is closely related to nitrogen supply, available moisture, and organic matter levels in the soil, and that these factors may be profoundly influenced by establishment practices.

Nitrogen and Soil Moisture

In the study reported by Squire et al. (1979), it was postulated that better early growth on 2R sites compared with matched 1R sites was due primarily to a mulch of litter and logging residue which, by reducing weed competition and evaporative losses, increased the availability of soil moisture. Furthermore, it was suggested that more favorable moisture levels in the surface soil on 2R sites increased nitrogen availability through accelerated mineralization of organic nitrogen. To test these hypotheses two further studies were initiated: one to examine the potential rates of ammonification and nitrification in the surface (0-3 cm) layer of Caroline sand (Hopmans et al. 1980), and one to evaluate the efficiency of a mulch in reducing evapotranspiration losses (Farrell, unpublished).

Nitrogen Mineralization in Caroline Sand

In the study reported by Hopmans et al. (1980), tests were conducted over a range of soil moisture contents (by weight) from 5% to 35% (soil water potential of -0.18 to -0.002 MPa; Squire, unpublished data) and incubation periods from 15 to 90 days at 30°C. Results showed that ammonification rates were relatively high and, as expected, strongly dependent on moisture content, with soil (C/N = 31) from under 27-year-old radiata pine producing up to 68 ppm ammonium at 15% soil moisture content after 60 days incubation compared with 30 ppm at 5% soil moisture content.

Squire (unpublished data) estimated that at 10 cm soil depth on matched 1R plots, moisture contents ranged from 3.4% to 5.6% (-0.12 to -0.86 MPa) over a four-month period during the first summer and autumn after planting. But on the 2R plots, moisture contents remained above 5.6% for all except one month of this period. In fact, subsequent measurements (Farrell, unpublished data) suggest that moisture contents may have been as high as 15% for part of this period. These results, supported by evidence of higher nitrogen levels in the foliage of trees from 2R than 1R plots during the first year after planting (Squire et al. 1979), reinforce the hypothesis that the litter and logging residue remaining after the 1R crop was felled increased nitrogen availability and contributed to enhanced 2R growth.

Litter and Logging Residue as a Mulch

A fundamental problem with the study reported by Squire et al. (1979) was that different weed populations developed on 1R and 2R sites, making it difficult to separate the effect of litter and logging residue in reducing soil evaporation and transpiration (suppressing weeds). Hence a further study was established in 1979 to measure these separate effects.

A factorial design with split plots was employed, with four major and two minor treatments replicated three times. The four major treatments were: (1) logging residue chipped in situ and spread uniformly over the undisturbed litter layer (M); (2) logging residue and litter layer completely removed from the site to bare soil with soil disturbance kept to a minimum (S); (3) as for treatment M plus a nonstimulatory chemical weed control treatment (MWc) (Weed control was by a combination of spraying with Tryquat and follow-up hand weeding, applied four times during the first year after planting. Residual herbicides were avoided because of potential growth modifying effects; for example, atrazine stimulation of radiata pine reported by Sands and Zed 1979); and (4) as for treatment S plus the weed control treatment described above (SWc).

Each plot was 24 m by 48 m (10 by 20 trees) and was split with ± fertilizer treatment (F). The fertilizer regime was a modified version of the maximum growth sequence described by Woods (1976). Height-graded radiata pine seedlings were planted (2.4 m by 2.4 m) between 1R stump lines during August 1979. The site formerly carried a mature, unthinned radiata pine stand of SQ IV planted in 1951 and clearcut during 1978. Preliminary growth results at age two years after planting are given in Table 2.

Table 2. Gains in mean diameter at 5 cm above ground level over control (treatment S, -F) at age two years.

Site Preparation Treatment	Diameter Gain (mm)		
	-F	+F	F alone
S	0.0a[1]	8.7a	8.7*
SWc	28.8b	51.1b	24.3***
M	17.1b	33.2c	16.1***
MWc	41.2c	55.0b	13.8**

[1]Values in each column followed by the same letter are not significantly different at $P < 0.05$.
*, **, *** significant at the 5%, 1%, and 0.1% levels respectively.

Diameter growth was significantly higher in treatments SWc, M, and MWc than in treatment S (control) irrespective of fertilizer treatment. A preliminary analysis of soil moisture data indicates that these responses were due primarily to reductions in evaporation or transpiration. Further analysis will enable these reductions to be quantified for each treatment. The response due to fertilizer was significant for all treatments, and a significant site preparation x fertilizer interaction was found ($P < 0.01$). Although the reasons for this interaction are not clear, it is apparent that fertilizer requirements are less when both the mulch and weed control treatments are applied.

Soil moisture data will ultimately indicate the effectiveness of the mulch in reducing the separate components of evaporation and transpiration. Nevertheless, the growth results in Table 2 support the mulch hypothesis and its ramifications advanced by Squire et al. (1979).

Organic Matter

Conservation of organic matter is considered to be a key factor in the maintenance of productivity on soils of relatively low fertility such as the sandy soils at Rennick. For such soils, organic matter levels have a significant effect on soil moisture availability and nutrient retention, as noted by Florence and Lamb (1975). Flinn et al. (1980) stressed that cultural practices such as burning and cultivation can rapidly deplete soil organic matter levels, which are then difficult to restore. In the study reported by Flinn et al. (1979), temperatures at the soil-litter interface in a mild broadcast burn ranged from <120°C to >670°C, with a mean of around 500°C. Although no quantitative measurements were made of soil organic matter before and after burning, it was evident from the bleached nature of the soil surface that losses occurred. A more complete combustion of organic matter in the surface soil would be expected under intense broadcast or windrow burns such as have been employed in the past in this locality.

Flinn et al. (1979) measured nutrient contents of litter and logging residue before and after burning, and these data have been used to construct a nitrogen budget for a radiata pine ecosystem (Table 3).

Table 3. A nitrogen budget for a 25-year-old unthinned, SQ V, 1R stand of radiata pine at Rennick.

Component	Nitrogen	
	(kg/ha)	(%)
Harvested wood (excluding bark)	203	9.1
Logging residue:		
wood plus bark	105	4.7
needles	114	5.1
Litter	204	9.1
Soil to 50 cm (excluding stumps and large roots)	1,610	72.0
Total	2,236	100.0

This budget indicates that nitrogen losses may vary from 9.1% where the residue is left in situ to 28% where all the residue is burned. Losses due to leaching down the soil profile, through soil and ash movement from the site by strong winds, and through direct combustion of soil organic matter are not considered because they were not measured by Flinn et al. (1979).

REFERENCES

Anon. 1978. Annual report, 1977-78. Woods and Forests Dept. S.A. 51 pp.

Bednall, B. H. 1968. The problem of lower volumes associated with second rotations in *Pinus radiata* plantations in South Australia. Woods and Forests Dept. S.A. Bull. 17.

Bengtson, G. W. 1978. Strategies for maintaining forest productivity: A researcher's perspective. In: T. Tippin, ed., Proceedings: A symposium on principles of maintaining productivity on prepared sites, pp. 123-159. Mississippi State University.

Evans, J. 1975. Two rotations of *Pinus patula* in the Usutu forest, Swaziland. Comm. For. Rev. 54(1):69-81.

Flinn, D. W. 1978. Comparison of establishment methods for *Pinus radiata* on a former *P. pinaster* site. Aust. For. 41(3):167-176.

Flinn, D. W., P. Hopmans, P. W. Farrell, and J. M. James. 1979. Nutrient loss from the burning of *Pinus radiata* logging residue. Aust. For. Res. 9:17-23.

Flinn, D. W., R. O. Squire, and P. W. Farrell. 1980. The role of organic matter in the maintenance of site productivity on sandy soils. N.Z. J. For. 52(2):229-236.

Florence, R. G., and D. Lamb. 1975. Ecosystem processes and the management of radiata pine forests on sand dunes in South Australia. Ecol. Soc. Aust. Proc. 9:34-48.

Hopmans, P., D. W. Flinn, and R. O. Squire. 1979. Soil chemical properties under eucalypt forest and radiata pine plantations on coastal sands. For. Comm. Vic. For. Tech. Pap. 27:15-20.

Hopmans, P., D. W. Flinn, and P. W. Farrell. 1980. Nitrogen mineralization in a sandy soil under native eucalypt forest and exotic pine plantations in relation to moisture content. Comm. Soil Sci. Plant Anal. 11(1):71-79.

Keeves, A. 1966. Some evidence of loss of productivity with successive rotations of *Pinus radiata* in the south-east of South Australia. Aust. For. 30(1):51-63.

Leitch, C. J., and P. W. Farrell. 1980. Evaluation of mechanical methods for treating *Pinus radiata* logging residue. In: Agricultural Engineering Conference 1980, Geelong, Victoria, pp. 21-26.

Lewis, N. B., A. Keeves, and J. W. Leech. 1976. Yield regulation in South Australian *Pinus radiata* plantations. Woods and Forests Dept. S.A. Bull. 23.

Ruiter, J. H. 1979. Forest water relations in south-eastern South Australia. CSIRO Div. For. Res. Annual report, 1978-79, pp. 20-26.

Sands, R., and G. D. Bowen. 1978. Compaction of sandy soils in radiata pine forests. II. Effects of compaction on root configuration and growth of radiata pine seedlings. Aust. For. Res. 8:163-170.

Sands, R. and P. G. Zed. 1979. Promotion of nutrient uptake and growth of radiata pine by atrazine. Aust. For. Res. 9:101-110.

Squire, R. O. 1982. Review of second rotation silviculture of *Pinus radiata* plantations in southern Australia: Establishment practice and expectations. Aust. For. 46(2):83-90.

Squire, R. O., D. W. Flinn, and P. W. Farrell. 1979. Productivity of first and second rotation stands of radiata pine on sandy soils. I. Site factors affecting early growth. Aust. For. 42(4):226-235.

Stephens, C. G., R. L. Crocker, B. Butler, and R. Smith. 1941. A soil and land use survey of the Hundreds of Riddoch, Hindmarsh, Grey, Young, and Nangwarry, County Grey, South Australia. CSIRO Aust. Bull. 142.

Switzer, G. L. 1978. Determinants of forest stand productivity. In: T. Tippin, ed., Proceedings: A symposium on principles of maintaining productivity on prepared sites, pp. 14-27. Mississippi State University.

Van Goor, C. P. 1952. Deep cultivation and the production capacity of dry sandy forest soils. Uitvoerige Verslagen 1:50-99.

Whyte, A. G. D. 1973. Productivity of first and second crops of *Pinus radiata* on the Moutere gravel soils of Nelson. N.Z. J. For. 18(1):87-103.

Woods, R. V. 1976. Early silviculture for upgrading productivity on marginal *Pinus radiata* sites in the south-eastern region of South Australia. Woods and Forests Dept. S.A. Bull. 24.

12

MONITORING THE LONG-TERM EFFECTS OF MANAGEMENT PRACTICES ON SITE PRODUCTIVITY IN SOUTH AFRICAN FORESTY

C. J. Schutz

INTRODUCTION

The productivity of successive conifer monocultures throughout the world has been the subject of many studies, recently reviewed by Evans (1976). The progress with research into the causes of second-rotation decline in productivity of _Pinus radiata_ in southeast Australia is described by Farrell et al. (1981). Studies by Evans (1975) have not indicated any loss in productivity in the second rotation of _P. patula_ pulpwood plantations in Swaziland.

In South Africa, existing monitoring techniques have revealed no evidence of any decline in site productivity after several rotations of pine, eucalypt, or wattle monocultures. Wattle (_Acacia mearnsii_) yields have, in some instances, increased after five rotations (A. P. G. Schönau, pers. comm.).

But increasing demands on the forest site by intensified management, coupled with potentially harmful processes operating in pine ecosystems (von Christen 1959, 1961, and 1964), could prove critical in maintaining site productivity in the long term. The influence of some of these factors is outlined below.

Because long-term effects clearly need to be monitored, a research program is under way (1) to determine whether changes in growth or yield over successive rotations are in fact taking place, (2) to determine whether changes, if any, are the result of management practices, changes in soil properties, or natural factors, and (3) to examine management systems that might eliminate or minimize any decline in growth or degrading of soil properties.

EFFECTS OF MANAGEMENT PRACTICES ON SITE PRODUCTIVITY

Nutrient Depletion

The loss of nutrients from plantation ecosystems through shortening the rotation or increasing the degree of utilization has recently been reviewed by Pritchett (1979) and Carlisle (1980).

The possibility of eventual nutrient depletion affecting yields in subsequent rotations of unthinned, short rotation, pulpwood plantations in South Africa is not so remote where these have been allocated to the poorest sites, particularly sandy soils. Replacement of nutrient losses through fertilizers may prove difficult (Bengtson 1978; Carlisle 1980).

Nutrient input via precipitation, aerosols, dust, nitrogen fixation, and mineral weathering is discussed by Carlisle. The improvement in the availability of nutrients to subsequent rotations on sites with rocky or dense subsoil through "fissuring" should also not be underestimated. This is the exploitation of decayed-root channels of the previous crop by roots in the present rotation (Figure 1).

Changes in Surface Soil Structure

Favorable soil structure may be adversely affected by the use of heavy mechanical equipment on certain soil types and under certain moisture conditions. This has been studied by Hatchell et al. (1970), but the role of soil texture is by no means clear. Many factors determine the recovery time for a compacted soil to regain its former structure.

Figure 1. Penetration of subsoil by new roots exploiting old root channels (*fissuring*)--*Pinus elliottii*, Tweefontein State Forest.

A fine, granular or crumb-structured surface horizon 50 to 100 mm thick has frequently been found to develop under pine stands in certain parts of South Africa (Figure 2). A similar structure is not found under adjoining unafforested vegetal cover. This improvement in structure is thought to be due to mycorrhizal activity (Grey 1981).

Loss of Topsoil

Erosion losses after the logging of steep slopes may be excessive. The choice of cropping practice can produce a 250-fold difference in soil loss (Nutter and Douglas 1978).

Fire may cause an even more severe soil loss. A slash disposal burn on a 10-degree slope after clearcutting *P. patula* in Swaziland resulted in a topsoil loss of about 16.5 t/ha within four months (Germishuizen and Badenhorst 1977). Broadcast burning of wattle brushwood in Natal led to the loss of 113 t/ha of topsoil in the first growing season (Sherry 1971). The crumb-structured surface horizon that develops under pines is particularly erodible. Wildfires and broadcast burning of slash on even moderate slopes in the eastern Transvaal have led to the almost total disappearance of this horizon over wide areas, exposing a hard, subsurface horizon resistant to moisture infiltration.

Figure 2. Loose, crumb-structured surface horizon attributed to mycorrhizal fungal action under a mature stand of *Pinus taeda*, Tweefontein State Forest.

Topsoil removal during mechanical site preparation is a problem in the southeastern United States (Pritchett and Wells 1978). It is unlikely to become a factor in South African forestry, since land clearing of this type is not necessary.

In contrast, severe gully erosion on granite soils in the northern and eastern Transvaal resulting from previous agricultural mismanagement has been stabilized after a single rotation of pine or *Eucalyptus*.

Fire

Fire may adversely affect many other soil properties, with effects directly related to the severity of the burn. Thus hazard reduction burns are likely to influence soil properties less than slash disposal burns or wildfires (Pritchett 1979). Farrell et al. (1981) have demonstrated the serious consequences for site productivity of organic matter destruction through burning practices on sandy soils in Australia.

Slash burning frequently results in the spread of the root rot fungus *Rhizina undulata* Fr., causing the death of newly planted *P. patula* (Schutz 1977).

Litter

The importance of the litter layer in plantation forestry, particularly on coarse-textured soils, has been clearly demonstrated by Squire et al. (1979) in the context of the second-rotation decline of *P. radiata* in Australia. They emphasize the role of litter in both moisture retention (mulching effect) and nutrient supply (mainly N). On sandy soils in regions of low rainfall in South Africa, site productivity could, as in Australia, be impaired by the removal of litter through burning or any other means.

On the other hand, excessive litter buildup creates problems of a different nature. Undecomposed litter layers up to 500 mm thick (± 320 t/ha oven dry) are accumulating under some first rotation stands of *P. patula* in the eastern Transvaal (Figure 3). The reason for the buildup of these mor layers is unknown. Preliminary results from a site-factor survey show little correlation with site index. A survey recently completed in a CCT (Correlated Curve Trend) spacing and thinning experiment indicated little variation in litter depth over stand densities ranging from 2,965 to 124 stems/ha, the average depth being 250 mm (Schutz et al., in preparation). A survey to determine the role of site factors in litter decomposition is currently under way.

Mor layers under *P. patula* are characterized by distinct L, F, and H layers, but the H layer (O_2 horizon) is usually less than 10 mm thick, or sometimes entirely absent, with the original preafforestation grasses still recognizable below. Rooting in the F layer is prolific, and this layer is obviously a source of moisture and nutrients, particularly on poorer sites. From this point of view the slow rate of decomposition is an advantage, but it may be outweighed by potentially harmful influences on the soil.

Figure 3. Undecomposed litter layer (0 to 500 mm) under a mature stand of *P. patula*, Ceylon State Forest.

Other Management Factors

Recent years have seen the appearance in South Africa of new pests, for example the black pine aphid (*Cinara cronartii* T and P). Indications are that some losses have been the direct result of overstocking (Bredenkamp 1979). Thinning-to-waste, however, provides favorable habitats for bark beetle. Retention of *P. taeda* stands beyond a rotation of forty years may result in decline by various agencies (Schutz and Wingfield 1979).

MONITORING PROCEDURES

The consequences of the increasing demands being made on the site by management practices clearly need to be monitored on a long-term basis. Site and yield changes over successive rotations may be investigated by using either paired plots on different sites or long-term plots on the same site (Farrell et al. 1981). Although the first method has the advantage of early results, it is a somewhat haphazard approach, the difficulty of matching sites exactly making its use limited and less precise than monitoring on the same site. With sufficient care to ensure duplication of silvicultural treatments over successive rotations and accurate recording of uncontrollable factors such as climate, the second method will be more reliable.

There is, however, a major problem associated with monitoring successive rotations: ensuring that each crop is genetically the same as its predecessor. Plots reestablished from seed collected from the previous crop would be acceptable provided no thinning had been applied, but isolation to prevent pollen contamination from surrounding stands, if these had not been established from the same seed source, presents almost insurmountable difficulties in ensuing rotations.

The solution to the problem is, however, quite simple and is being applied in the long-term monitoring program in South Africa. The seed used originates from specific crosses performed in clonal banks at the end of each rotation. The clones selected will be retained indefinitely and rejuvenated whenever necessary by fresh grafted material. The crosses selected are those used as a "genetic check" in progeny testing, thus ensuring comparability with successive generations in the breeding program.

Although the paired plot system will be used where feasible to provide supplementary information, as was done by Squire et al. (1979), permanent monitoring plots are being established on a wide variety of sites.

Single-treatment (standard silviculture), replicated plot trials are planned for some sites, but on others various treatments will be applied. On low-quality sites a fertilizer treatment will be added in order to determine whether depleted nutrients could be replaced by this means. In short rotation, unthinned, pulpwood plantations a whole-tree utilization treatment will be included. The long-term effects of retaining, reducing, or removing thick litter layers will be monitored in another series. Although some of these treatments can be expected to have some effect within one rotation, they are to be repeated in successive rotations, while all standard silvicultural treatments (espacement, thinnings, prunings, etc.) will be duplicated exactly. Although uncontrolled events (drought, fire, hail, pests) cannot be duplicated, their effects can be monitored, for example through stem analyses, and adjustments can be made.

Periodic growth measurements and volume yields from thinnings plus clearcuttings (or biomass) will be the criteria by which productivity will be judged, and soil changes will be monitored.

Soil and nutrient losses through erosion after clearcutting and other aspects of nutrient cycling are being investigated in gauged catchments at several research stations.

REFERENCES

Bengtson, G. W. 1978. Strategies for maintaining forest productivity: A researcher's perspective. In: T. Tippin, ed., Proceedings: A symposium on principles of maintaining productivity on prepared sites, pp. 123-159. Mississippi State University.

Bredenkamp, B. V. 1979. Thin or lose all. Forestry News 3/79. Dept. of Forestry, Pretoria, South Africa.

Carlisle, A. 1980. Forest exploitation and environmental objectives. Eleventh Commonwealth Forestry Congress, Trinidad.

Evans, J. 1975. Two rotations of Pinus patula in the Usutu forest, Swaziland. Comm. For. Rev. 54(1):69-81.

_________. 1976. Plantations: Productivity and prospects. Aus. For. 39(3):150-163.

Farrell, P. W., D. W. Flinn, R. O. Squire, and F. G. Craig. 1981. On the maintenance of productivity of radiata pine monocultures on sandy soils in southeast Australia. XVII IUFRO World Congress, Japan, Congress Group S1.02, Division 1, pp. 117-128.

Germishuizen, P. J., and C. J. Badenhorst. 1977. The influence of harvesting on silviculture. S.A. For. J. 102:29-36.

Grey, D. C. 1981. New trends in soil classificiation. S.A. Institute of Forestry, Soils course. Unpublished.

Hatchell, G. E., C. W. Ralston, and R. R. Foil. 1970. Soil disturbances in logging: Effects on soil characteristics and growth of loblolly pine in the Atlantic Coastal Plain. J. For. 68(12):772-775.

Nutter, W. L., and J. E. Douglas. 1978. Consequences of harvesting and site preparation in the Piedmont. In: T. Tippin, ed., Proceedings: A symposium on principles of maintaining productivity on prepared sites, pp. 65-72. Mississippi State University.

Pritchett, W. L. 1979. Properties and management of forest soils. John Wiley and Sons, New York. 500 pp.

Pritchett, W. L., and C. G. Wells. 1978. Harvesting and site preparation increase nutrient mobilization. In: T. Tippin, ed., Proceedings: A symposium on principles of maintaining productivity on prepared sites, pp. 98-110. Mississippi State University.

Schutz, C. J. 1977. Rhizina undulata: 'n Gevaarlike Kalant. Forestry News 4/77. Dept. of Forestry, Pretoria, South Africa.

Schutz, C. J., and M. J. Wingfield. 1979. A health problem in mature stands of Pinus taeda in the Eastern Transvaal. S.A. For. J. 109:47-49.

Schutz, C. J., B. V. Bredenkamp, and M. A. Herbert. In preparation. The influence of stand density on litter depth in the Nelshoogte *Pinus patula* C.C.T.'s.

Sherry, S. P. 1971. The black wattle. University of Natal Press.

Squire, R. O., D. W. Flinn, and P. W. Farrell. 1979. Productivity of first and second rotation stands of radiata pine on sandy soils. I. Site factors affecting early growth. Aust. For. 42(4):226-235.

von Christen, H. C. 1959. The forest soils of the Transvaal Mistbelt. Dept. of Agriculture and Technical Services, Div. Chem. Services Report 1096/59. Unpublished.

__________. 1961. The soils and ecology of the pine plantations of the Kaapsehoop Plateau. Unpublished report.

__________. 1964. Some observations on the forest soils of South Africa. Forestry in South Africa No. 5.

13

EXPERIENCES WITH STEM PHLOEM ANALYSIS

M. Alcubilla and K. E. Rehfuess

INTRODUCTION

During the past decades foliar analysis has proved to be an effective approach for evaluating the nutritional status of forest stands, provided sufficient representative trees are examined and a standardized sampling technique is used. In old and dense stands, however, sampling leaves from the uppermost section of the standing crown is technically difficult, expensive, and risks damaging the trees seriously. Therefore, the use of other tissues or organs for diagnostic purposes should be considered. Stimulated by some earlier reports (Bolle-Jones 1957; Will 1962; White et al. 1970, 1972; van den Driessche 1974; Alcubilla and Rehfuess 1975; Parusheva and Evanglatova 1975; Hohenadl et al. 1978) we have carried out some pilot investigations with stem phloem (inner bark) to find out whether the mineral contents of this tissue allow us to judge the supply of nutrients in trees.

MATERIAL

In different years between 1974 and 1979, seventy-five stands were included in this program, namely thirty-four Norway spruce (*Picea abies*), eighteen Scots pine (*Pinus sylvestris*), five silver fir (*Abies alba*), and eighteen oak (*Quercus petraea*) forests. Sixty-two stands were part of experiments about the effects of fertilization and sewage sludge application. Thirteen stands were investigated in the course of inventories in unfertilized forests, widely differing in their nutritional status. The sample stands covered a great variety of parent rocks and soils in southern Germany.

Current (half-year-old) foliage of spruce and pine was sampled between October and December from the topmost whorl, and oak leaves in August from the upper crown. Samples of silver fir were taken at the onset of the growth period (June), since the experiment was concerned with other questions as well.

More or less simultaneously, four phloem disks of 15 mm diameter were cut from the trunk by means of a cork borer either at its base (young trees) or at breast height (older trees). The wounds were covered with wax. These phloem disks were carefully cleaned of dead bark. In most cases ten trees per stand were sampled for both foliage and phloem. The individual samples were integrated into one composite sample per stand and analyzed by standard procedures. To examine the tree-to-tree variation, twenty phloem disks per tree were taken in some of the experiments.

All analytical data are given on a dry-matter (65°C) basis.

RESULTS

Tissue Nutrient Levels

The nitrogen, phosphorus, potassium, and iron contents were generally higher in foliage than in phloem. The latter, on the other hand, was normally richer in calcium. The magnesium, copper, and manganese contents in both sets of samples varied on similar levels. Norway spruce phloem proved to be extremely rich in zinc (Tables 1-3).

In most cases nitrogen, phosphorus, calcium, and manganese in foliage and phloem were positively correlated, provided the experimental manipulation or the site factors had led to sufficient variation of the element levels in both tissues. Examples of a lack of correlation are Norway spruce in Expt. Hilpertsberg (for N) and Scots pine in Expt. Burglengenfeld (for Ca), where fertilization had not altered significantly the nitrogen and calcium levels in either needles or phloem.

The experimental manipulations (fertilization and sewage sludge disposal) affected the nitrogen, phosphorus, calcium, and manganese levels in leaves and inner bark mostly in the same way. Site differences or treatment effects with regard to potassium, magnesium, copper, zinc, and iron in both tissues generally were small. This may be why we detected significant (positive) correlations for the contents of these elements in leaves and phloem only in a few cases.

In each of five silver fir stands, we investigated ten healthy trees and ten trees affected by the so-called silver fir disease. Pooling all analyses without taking into account the occurrence and intensity of the disease, only the calcium and manganese contents of needles and phloem were significantly correlated, but not the nitrogen and phosphorus levels as in most other tests. This deviation from the normal findings may be due either to the impact of the disease or to the unusual material and date of sampling (one-year-old needles in June).

By means of multiple discriminant analysis, we computed for some experiments and for each individual plot the accuracy of correct classification into the different treatments, using the macronutrient or micronutrient contents of needles or phloem as parameters. The principal idea underlying this approach was: the more the nutrient levels in the tissues vary characteristically as dependent on treatment, the higher should be the degree of correct classification by this statistical procedure. We were astonished to see that in many cases phleom nutrient contents produced better classification results compared with foliage levels.

Figure 1 gives a geometric interpretation of discriminant analysis results for one of the trials. The position of each plot (letters) and of treatment means (numbers) is indicated on a so-called hyperplain, which allows an optimal separation of treatments. This position is determined by two canonical variables, which are computed by means of canonical correlation analysis as two statistically independent, optimal, and linear combinations of needle or phloem parameters. For Scots pine and the Expt. Burglengenfeld, and considering the macronutrient levels only, the separation of treatment means was better using phloem than foliage characteristics.

Table 1. Range of nutrient contents in foliage (L) and phloem (B) of Norway spruce (*Picea abies*) and interrelations of tissue nutrient levels and growth parameters.

Type and Year of Investigation, Number of Plots (stand age), Correlated Parameters		Macronutrients (mg/g) N	P	K	Ca	Mg	Micronutrients (ug/g) Mn	Fe	Cu	Zn
		r-values for different sets of parameters					r-values for different sets of p.			
Inventory in the Baar-Wutach	L	9.9-14.7	1.0-1.6	6.1-7.6	2.5-7.0	1.0-1.8	129-4,055			
region, 1969-78	B	2.6-4.4	0.5-0.7	3.2-4.3	4.3-24.2	0.5-0.8	81-3,076			
n=8 (65-83 years)										
Nutrient in L/Nutr. in B		.92*	.54	.00	.93*	-.02	.99*		n.d.	
Nutrient in L/i_v**		.97*	.63	-.63	.05	.54	.15			
Nutrient in B/i_v**		.88*	.83*	.05	.08	.66	.19			
Fertilization Experiment	L	11.3-14.2	1.5-2.1	6.6-8.8	4.6-9.7	0.7-1.2	22-55	30-56	7-35	30-53
Gachenwinkel, 1975	B	3.2-5.5	0.5-0.7	3.8-4.8	13.2-18.9	0.8-1.1	27-62	14-25	15-22	163-189
n=9 (23-27 years)										
Nutrient in L/Nutr. in B		.81*	.64*	.57	.53	-.44	.88*	-.49	-.62*	.06
Nutrient in L/i_h**		.76*	-.26	.70*	-.68*	-.56	.77*	.62	.40	-.38
Nutrient in B/i_h**		.65*	-.13	.45	-.82*	-.47	.92*	-.72*	-.22	-.29
Fertilization Experiment	L	13.1-16.3	1.2-1.9	7.5-9.4	2.2-6.7	0.4-1.7	1,156-2,469	81-125	7-13	27-47
Hilpertsberg, 1975	B	4.1-5.0	0.5-0.8	3.4-4.8	6.1-10.1	0.8-1.4	1,099-3,434	26-70	10-15	143-193
n=11 (26 years)										
Nutrient in L/Nutr. in B		.00	.70*	-.05	.70*	.61*	.66*	.29	-.22	-.36
Sewage sludge Experiment	L	11.7-14.4	1.5-1.8	5.1-6.7	3.1-4.3	0.8-1.1	1,427-2,227	42-58	4-9	20-31
Hallerholz, 1975	B	3.2-4.6	0.6-0.8	3.4-4.3	7.1-10.7	1.5-1.6	1,180-1,884	28-43	9-11	113-131
n=6 (70 years)										
Nutrient in L/Nutr. in B		.72	.52	-.49	-.71	-.58	.45	.91*	.41	-.60

* r significant at the 95% level.

** i_v = mean annual volume increment" i_h = height increment.

Table 2. Range of nutrient contents in foliage (L) and phloem (B) of Scots pine (*Pinus sylvestris*) and interrelations of tissue nutrient levels and growth parameters.

Type and Year of Investigation, Number of Plots (stand age), Correlated Parameters		Macronutrients (mg/g) N	P	K	Ca	Mg	Micronutrients (µg/g) Mn	Fe	Cu	Zn
		r-values for different sets of parameters					r-values for different sets of p.			
Fertilization Experiment Waldsassen, 1975	L	13.0-16.9	1.5-2.0	5.8-6.2	2.2-3.6	0.6-0.9	238-650	94-139	7-10	50-63
n=9 (97 years)	B	4.3-5.7	0.9-1.0	4.9-5.6	8.0-16.0	1.5-1.7	156-575	45-81	9-11	35-46
Nutrient in L/Nutr. in B		.80*	.66*	.47	.85*	.39	.96*	.21	-.52	.75*
Nutrient in L/i_v**		.87*	.63	.03	.74*	.70*	-.37	-.01	-.13	.32
Nutrient in B/i_v**		.80*	.81*	.39	.92*	.66	-.43	-.33	.55	-.40
Fertilization Experiment Burglengenfeld, 1975	L	13.8-17.2	1.4-1.7	4.7-5.5	3.0-4.0	0.7-1.0	460-1,110	80-135	5-7	36-49
n=9 (88 years)	B	3.7-5.4	0.7-1.1	4.0-5.5	13.1-29.1	1.7-2.0	380-842	43-59	8-28	37-60
Nutrient in L/Nutr. in B		.92*	.85*	.73*	-.28	-.59*	.90*	-.53	.24	.37
Nutrient in L/i_v**		.83*	.85*	.69*	.09	-.42	-.89*	-.05	-.06	-.16
Nutrient in B/i_v**		.89*	.92*	.67	.80*	.27	-.93*	-.07	.06	-.68

* r significant at the 95% level.
** i_v = mean annual volume increment.

Table 3. Range of nutrient contents in foliage (L) and phloem (B) of oak and silver fir and interrelations of tissue nutrient levels.

Type and Year of Investigation, Number of Plots (stand age), Correlated Parameters		Macronutrients (mg/g) N	P	K	Ca	Mg	Micronutrients (µg/g) Mn	Fe
		r-values for different sets of parameters					r-values	
Oak (Quercus petraea)								
Fertilization Experiment Rothenbuch, 1978	L	21.3-27.6	1.6-2.6	8.0-11.4	5.1-8.8	0.9-2.2	2,050-4,390	94-140
n=8 (23 years)	B	3.6-5.2	0.3-0.4	2.9-3.6	13.9-27.4	0.2-0.7	1,290-2,000	29-116
Nutrient in L/Nutr. in B		.66*	.65*	.53*	.57*	.73*	.42	.31
Silver fir (Abies alba)*								
Inventory Bavaria, 1978-79	L	11.8-18.6	1.1-1.4	3.8-5.6	3.1-17.0	0.7-1.7	840-6,044	52-143
n=5 (87-123 years)	B	2.6-3.8	0.3-0.4	1.9-2.5	4.1-14.1	0.2-0.6	379-1,470	19-60
Nutrient in L/Nutr. in B		-.05	-.05	.08	.92*	.40	.96*	-.33

* r significant at the 95% level.
** One-year-old needles and phloem sampled during June.

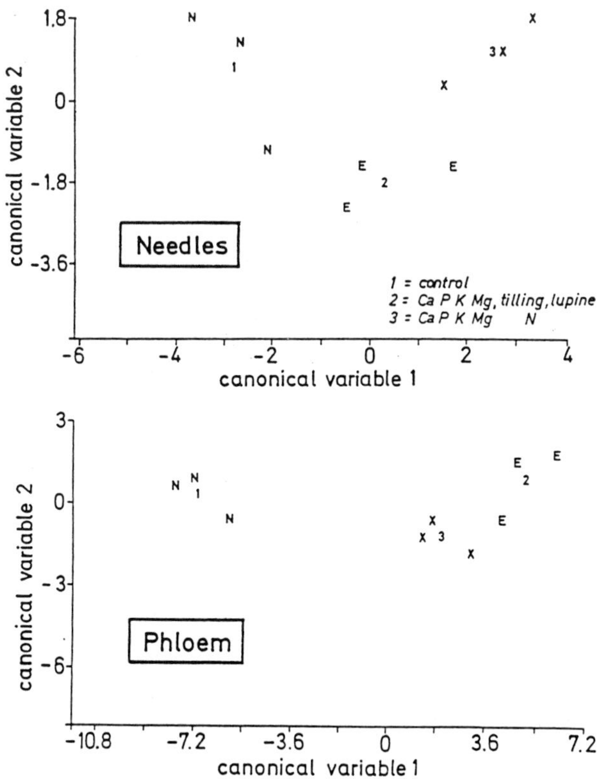

Figure 1. Separation of the treatments by discriminant analyses using the macroelement contents of needles and phloem (Pine Experiment Pustert).

Interrelations of Growth Parameters and Tissue Nutrient Levels

For those inventories and experiments where growth parameters could be related to nutrient contents in foliage and phloem, the statistical calculation produced for nitrogen and phosphorus correlation coefficients (r-values) of the same order of magnitude (Tables 1-3). The correlations were positive and (nearly) significant in all comparisons, with the one exception of phosphorus in the Norway spruce Expt. Gachenwinkel (Table 1). This result had to be expected, since the trees on this site were well supplied with phosphorus: the levels in both foliage and phloem varied on a rather high level more or less unaffected by fertilization.

For thirty-four spruce and eighteen pine stands of similar age, the data were pooled and single linear regressions for the nitrogen contents of both tissues were calculated. Using the rather well established critical levels for nitrogen deficiency in pine and spruce needles (15 and 13 mg N/g dry matter), a preliminary threshold value for nitrogen in stem phloem was derived. These regressions suggest pine and spruce trees suffer from nitrogen deficiency if the nitrogen contents in phloem fall below 4.5 and 4.1 mg/g dry matter, respectively.

Additional Observations with Regard to Methodological Problems

In our comparisons, phloem parameters generally showed a larger tree-to-tree variation than foliage characteristics. Analyzing two spruce trees, which were partitioned into five sections each, we found the nitrogen, phosphorus, and potassium contents of phloem increasing versus stem top, whereas the calcium contents decreased in the same direction (Figure 2). For magnesium, zinc, iron, and manganese no

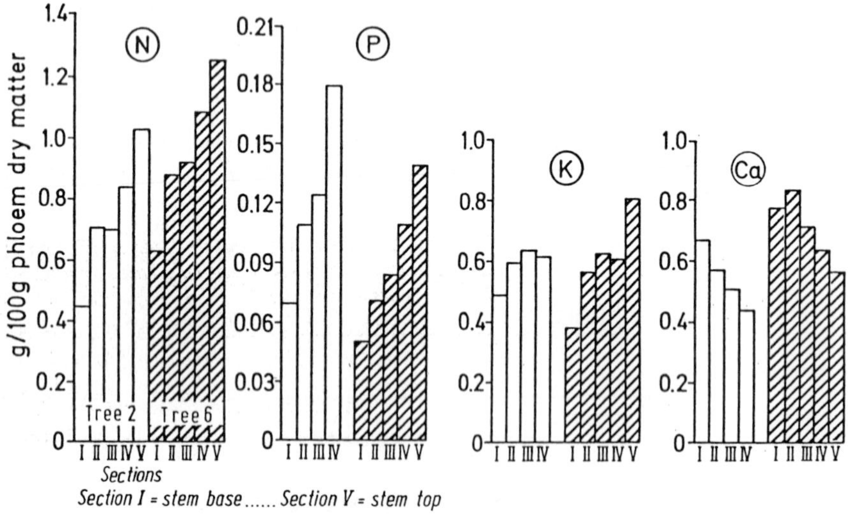

Figure 2. Nutrient contents of phloem as dependent on stem height (Spruce Experiment Hilpertsberg).

trends showed up. There is evidence that distinct stem positions should be preferred for sampling; the optimum position still has to be defined. In the spruce experiment with sewage sludge application, we investigated the seasonal variation of phloem nutrient levels by sampling monthly ten trees per plot. Although the data revealed a distinct sludge effect on nitrogen, phosphorus, and magnesium contents, no clear seasonal trend could be detected (Figure 3).

CONCLUSIONS

The details of a standardized and effective phloem analysis procedure still remain to be studied more intensively. Nevertheless, our preliminary observations give evidence that the degree of supply with nitrogen, phosphorus, calcium, and magnesium can be evaluated by phloem analysis with similar results and accuracy as with foliar analysis. Therefore, it seems worthwhile to follow this approach. We do

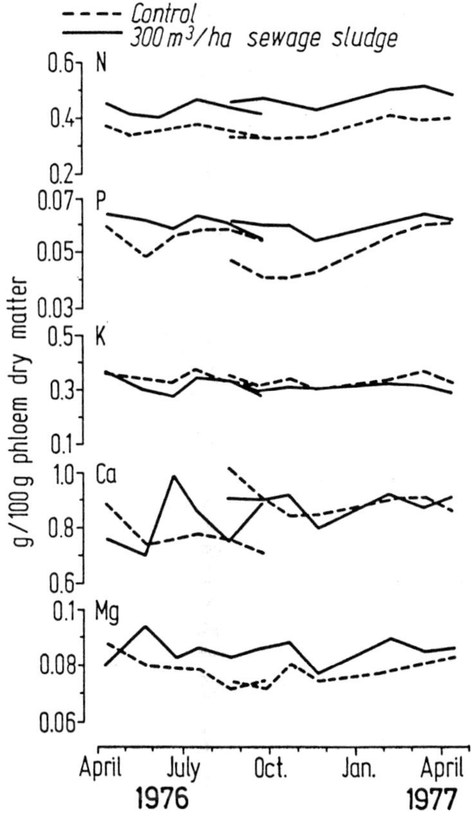

Figure 3. Seasonal variation of N, P, K, Ca, and Mg contents of stem phloem (Spruce Experiment Hallerholz).

not expect phloem analysis to replace foliar analysis, since the bulk of experiences for the latter technique is already enormous. Phloem analysis, however, presumably could be developed to an appropriate alternative in certain cases, such as for old, high, and dense stands, for circumstances in which the leaves are influenced by insects or fungi, and for periods when it is impossible to sample representative foliage. The disadvantage of a larger tree-to-tree variation of phloem nutrient contents may be overcome by better standardization of sampling.

REFERENCES

Alcubilla, M., and K. E. Rehfuess. 1975. Voruntersuchungen über die Eignung der Bastanalyse zur Beurteilung des Ernährungszustandes von Fichten. Forstw. Cbl. 94:344-351.

Bolle-Jones, E. W. 1957. Foliar diagnosis of mineral status of Hevea in relation to bark analysis. J. Rubb. Res. Inst., Malaya 15:109-127.

Hohenadl, R., M. Alcubilla, and K. E. Rehfuess. 1978. Die Stammbastanalyse als Methode zur Beurteilung des Ernährungszustands von Coniferen. Z. Pflanzenernähr. Bodenk. 141:687-704.

Parusheva, A., and N. Evanglatova. 1975. The content of N and some ash elements in the foliage and stem of *Pinus peuce*. Gorsko Stopanstvo 20:37-40.

van den Driessche, R. 1974. Prediction of mineral nutrient status of trees by foliar analysis. Bot. Rev. 40:347-397.

White, J. D., L. T. Alexander, and E. W. Clark. 1972. Fluctuations in the inorganic constituents of inner bark of loblolly pine with season and soil series. Can. J. Bot. 50:1287-1293.

White, J. D., C. G. Wells, and E. W. Clark. 1970. Variations in the inorganic composition of inner bark and needles of loblolly pine with tree height and soil series. Can. J. Bot. 48:1079-1083.

Will, G. M. 1962. Assessment of forest site capacity in New Zealand. In: G. J. Neale, ed., Transactions of the Joint Meeting Commission IV and V, International Society of Soil Science, pp. 803-807. Palmerston North, New Zealand.

14

ESTIMATION OF VOLUME AND WEIGHT GROWTH IN *GMELINA ARBOREA* WITH X-RAY DENSITOMETRY

A. E. Akachuku

INTRODUCTION

Gmelina arborea Roxb., a hardwood species of the family of Verbenaceae, is a native of Southeast Asia that has been introduced in many tropical and subtropical countries as a source of pulp. Because of the significance of this species in afforestation programs of many countries, it is important to study its growth rate in sites to obtain information on its probable yield in similar sites for a specified period of growth. High rate of growth is generally reported in the early years, and current annual increment (CAI) usually culminates between seven and nine years, after which it declines in unthinned plantations. Mean annual increment (MAI) reaches a maximum between eleven and fifteen years, depending on site.

In Nigeria the following rates of growth have been reported:

1. In the Derived Savanna Zone:
   (a) On poor sandy soils 84 m^3/ha (cubic meters per hectare) after twelve years' growth.
   (b) On good clay or laterite soils 210 m^3/ha after twelve years' growth.
   (c) On the most favorable savanna sites 252 m^3/ha after ten years' growth.
2. In the Rain Forest Zone: Little information is available but 252 m^3/ha after eight years' growth is reported.

From these data it is clear that a mean annual increment of 7 to 25.2 m^3/ha can be expected for the range of savanna sites and 31.5 m^3/ha for high forest (Chittenden et al. 1964).

A more recent study on site productivity of *Gmelina* plantations in moist evergreen forest and savanna area of Bendel (former Mid-Western) State of Nigeria has shown that unthinned plantations would be expected to produce 291 m^3/ha in fifteen years, this being when the MAI culminates at 19.4 m^3/ha. Maximum CAI of 28.6 m^3/ha is reached in eight years with a total standing volume of 127 m^3/ha (Greaves 1972, 1975).

In Malaysia the following MAI per hectare have been reported: 36.5 m^3 in seven years, 38.0 m^3 in eight years, 31.0 m^3 in nine years, and 28.8 m^3 in eleven years (Freezaillah and Sandrasegaran 1966).

In the Philippines, *Gmelina* can yield about 108 m^3/ha in three years, that is, MAI of 36 m^3/ha (Chinte 1971).

There is no information on weight yield of *Gmelina* plantations. Such information is necessary, since pulp mills express pulpwood in weight units. The objective of this study was to determine the volume and weight CAI, MAI, and cumulative yield in *Gmelina* pulpwood plantations in high forest and secondary regrowth forest areas in Nigeria.

MATERIALS AND METHODS

Volume and weight increments of Gmelina arborea were obtained from stem analysis. The procedure involved the counting of the growth rings present in the Gmelina wood and measuring the widths and densities (specific gravities) of individual rings. The sampling strategy was as follows:

Two sample plots (20 m x 20 m) were selected at random from seven-year-old Gmelina plantations in each of the following areas in Nigeria: (1) Ajebandele in Omo Forest Reserve, a high forest area in Ogun State (site 1); (2) Awi in Oban Forest Reserve, a secondary regrowth forest area in Cross River State (site 2). Five trees were selected at random from each sample plot. The mean volume of the five trees determined from the measurements of tree total height, diameter at breast height, and form factor was close to the mean volume of all the trees in a plot determined in the same way. This indicates that the sample was a good representation of the plot. Stem diameter at various heights was also measured for stem profile. Billets about 7.5 cm thick were cut from 5%, 25%, 45%, and 65% of the total height of each of the twenty sample trees. Two radial strips were cut from each billet for the determination of wood densities and widths of growth rings.

Measurement of Wood Density

Wood density was determined with X-ray densitometric technique (Hughes and Sardinha 1975). The procedure was as follows: (1) machining the two radial strips cut from each billet to accurate sizes; (2) conditioning them to equilibrium moisture content corresponding to laboratory conditions (about 12%); (3) producing negative images of the wood samples on X-ray films; (4) developing the films; and (5) scanning the radiographs on a microdensitometer that converts the film images to numerical values of density.

Estimation of Growth Sheath Volume and Weight

The width of each growth ring was obtained from the density values (both numerical values and graphs) from the densitometer. The width of each ring was the distance between the beginning earlywood and the end of latewood. The ring width values were converted to ring areas and plotted against their corresponding heights above the ground on graph paper to show the variation in the area of each growth ring along the tree bole. The volume of wood formed by the vascular cambium in the main bole in each year was computed and multiplied by its wood density to obtain its air-dry weight. From the values of individual trees, the CAI, MAI, and cumulative yield per hectare were computed and expressed in cubic meters and metric tons at air-dry moisture content (about 12%). The volume and weight excluded the values for the bark, branches, pith, and roots.

The values of the increment and yield were regressed on age, and the coefficient of determination, r^2, and the index of determination, i^2, were computed. These terms (r^2 and i^2) show what proportion of the variation in the values of the dependent variable can be explained by, or estimated from, the concomitant variation in the values of the independent variable. The term r^2 was used to describe linear relationships, while nonlinear relationships were described by i^2 (Ezekiel and Fox 1961).

RESULTS

Current Annual Increment (CAI)

In the high forest area (site 1) the volume CAI of the *Gmelina* plantations reached a maximum of 62.3 m^3/ha in the fourth year of growth and decreased to 42.7 m^3/ha in the seventh year. The weight CAI reached a maximum of 28.8 metric tons/ha in the fifth year and decreased to 23.5 in the seventh year (Table 1 and Figures 1 and 3).

In the secondary regrowth forest area (site 2), the volume CAI reached a maximum of 56.5 m^3/ha in the fifth year and decreased to 31.4 m^3/ha in the seventh year. The weight CAI reached a maximum value of 26.1 metric tons/ha in the sixth year and decreased to 21.2 in the seventh year (Table 2 and Figures 2 and 4).

Mean Annual Increment (MAI)

In seven years the volume and weight MAIs were 44.0 m^3/ha and 21.0 metric tons/ha respectively in site 1, and 33.0 m^3/ha and 16.3 metric tons/ha respectively in site 2 (Tables 1 and 2 and Figures 1 to 4).

Table 1. Volume and weight growth of *Gmelina* trees in high forest area.

		Volume (m^3/ha)			Weight (metric tons/ha)		
Sample	Age (years)	CAI	MAI	Cum. Growth	CAI	MAI	Cum. Growth
Plot 1	1	5.5	5.5	5.5	1.7	1.7	1.7
	2	40.8	23.1	46.3	17.0	9.4	18.7
	3	54.1	33.7	100.4	23.2	14.0	41.9
	4	71.3	42.9	171.7	32.4	18.6	74.2
	5	61.9	46.7	233.6	30.9	21.0	105.1
	6	59.6	48.8	292.8	32.5	22.9	137.6
	7	42.3	47.8	335.1	23.7	23.0	161.3
Plot 2	1	3.1	3.1	3.1	1.6	1.6	1.6
	2	25.9	14.5	29.0	11.5	6.6	13.2
	3	51.0	26.7	80.0	22.5	11.9	35.7
	4	53.3	33.3	133.3	24.9	15.1	60.6
	5	58.8	38.4	192.1	26.8	17.5	87.4
	6	46.3	39.7	238.4	22.7	18.3	110.1
	7	43.1	40.2	281.5	23.4	19.1	133.5
Site Mean	1	4.3	4.3	4.3	1.6	1.6	1.6
	2	33.3	18.8	37.6	14.2	8.0	15.9
	3	52.5	30.0	90.1	22.8	12.9	38.8
	4	62.3	38.1	152.4	28.6	16.8	67.4
	5	60.3	42.5	212.7	28.8	19.2	96.2
	6	52.9	44.2	265.6	27.6	20.6	123.8
	7	42.7	44.0	308.3	23.5	21.0	147.4

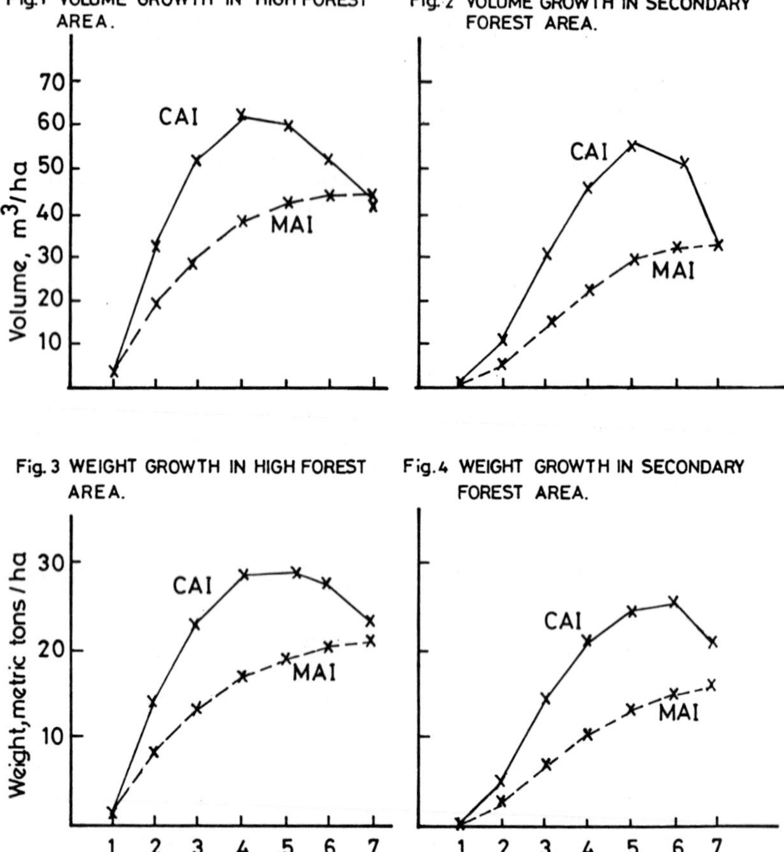

Variation in Gmelina growth rate (CAI and MAI) with age.

Table 2. Volume and weight growth of Gmelina trees in secondary forest area.

Sample	Age (years)	Volume (m^3/ha)			Weight (metric tons/ha)		
		CAI	MAI	Cum. Growth	CAI	MAI	Cum. Growth
Plot 1	1	1.1	1.1	1.1	0.2	0.2	0.2
	2	11.1	6.1	12.2	5.3	2.7	5.5
	3	30.3	14.1	42.2	13.5	6.3	19.0
	4	50.4	23.2	92.9	23.4	10.6	42.4
	5	52.6	29.1	145.5	23.0	13.1	65.4
	6	48.1	32.3	193.6	24.0	14.9	89.4
	7	34.7	32.6	228.3	19.3	15.5	108.7
Plot 2	1	1.6	1.6	1.6	0.7	0.7	0.7
	2	10.2	5.9	11.8	6.0	3.3	6.7
	3	36.1	16.0	47.9	16.0	7.6	22.7
	4	42.3	22.5	90.2	19.7	10.6	42.4
	5	60.4	30.1	150.6	26.9	13.8	69.3
	6	54.9	34.2	205.5	28.2	16.2	97.4
	7	28.2	33.4	233.7	23.2	17.2	120.6
Site Mean	1	1.3	1.3	1.3	0.4	0.4	0.4
	2	10.6	5.6	11.9	5.6	3.0	6.1
	3	33.2	15.0	45.1	14.7	6.9	20.8
	4	46.3	22.8	91.4	21.5	10.6	42.4
	5	56.5	29.6	147.9	24.9	13.4	67.3
	6	51.5	33.2	199.4	26.1	15.5	93.4
	7	31.4	33.0	230.8	21.2	16.3	114.6

Cumulative Growth

The volume and weight cumulative growth reached 308.3 m^3/ha and 147.4 metric tons/ha respectively in site 1, and 230.8 m^3/ha and 114.6 metric tons/ha respectively in site 2 (Tables 1 and 2).

Regression Analysis

It was observed that variation in CAI with age did not closely fit the linear model: r^2 ranged from 17% to 48% for the volume CAI and 44% to 78% for the weight CAI. It fitted the quadratic model very closely (80% to 99%). MAI and cumulative yield fitted the linear and quadratic models (r^2 and i^2 ranged from 80% to 99%), but the latter gave a better fit (Table 3). It is clear that a large proportion of the variation in CAI, MAI, or cumulative yield was associated with concomitant variation in age. If site conditions are fairly constant, the cumulative yield of Gmelina plantations will be expected to increase considerably with age in a short rotation.

Table 3. Variation in CAI, MAI, and cumulative yield with age: coefficient and index of determination (r^2 and i^2) as percentages.

	Volume Growth			Weight Growth		
Data	CAI	MAI	Cumulative yield	CAI	MAI	Cumulative yield
(1) Model: $Y = a + bx$						
Site 1, Plot 1	17	80	99	44	88	99
Site 2, Plot 2	33	85	99	52	88	99
Site 1 (Mean)	26	82	99	50	88	99
Site 2, Plot 1	48	93	97	65	96	97
Site 2, Plot 2	36	93	96	78	97	96
Site 2 (Mean)	43	93	97	73	96	96
(2) Model: $Y = a + bx + cx^2$						
Site 1, Plot 1	95	99	99	96	99	99
Site 1, Plot 2	94	99	99	94	99	99
Site 1 (Mean)	98	99	99	99	99	99
Site 2, Plot 1	90	97	98	92	98	98
Site 2, Plot 2	80	96	97	94	99	99
Site 2 (Mean)	87	97	98	94	98	99

DISCUSSION AND CONCLUSIONS

Although *Gmelina* grows fast in these sites within seven years, it is not known for how many cycles the sites can produce pulpwood at such a rate before the soils deteriorate. Forest management in these areas should include methods of improving soil fertility if the amount of wood to be produced after a number of rotations is to be close to the expected value.

The rotation age of about seven years chosen for the *Gmelina* pulpwood plantations in these areas was based not only on the usual relationship between CAI and MAI (Assmann 1970) but also on the average values of wood properties (such as density, fiber morphology, and chemistry) which control pulp and paper properties. The wood density was found to vary within and between trees. Therefore, equal volumes of wood from different sources may have different weights. Expressing wood in volume may be of little value to the wood user unless the density range is also stated. In a short rotation, the yield of forest plantations should be expressed in weight, because this gives a better indication of the proportion of cell wall than volume.

REFERENCES

Assmann, E. 1970. The principles of forest yield study. Pergamon Press, Oxford, New York. 506 pp.

Chinte, F. O. 1971. Silvicultural studies of four pulpwood species. Philipp. Lumberm. 17(5):8-26 and 29.

Chittenden, A. E., D. G. Coursey, and J. O. Rotibi. 1964. Paper making trials with Gmelina arborea timber in Nigeria. Tappi 47(12):186A-192A.

Ezekiel, M., and K. A. Fox. 1961. Methods of correlation and regression analysis. John Wiley and Sons, New York.

Freezaillah, C. Y., and K. Sandrasegaran. 1966. Growth and yield of Yemane Gmelina arborea Roxb. Malay Forester 29(3):140-153.

Greaves, A. 1972. Summary report on the results of the site and associated productivity studies of plantations of Gmelina arborea Roxb. in the (former) Mid-Western State of Nigeria. Dept. of Forestry and Wood Science, University College of North Wales, Bangor. Unpublished. 19 pp.

__________. 1975. Some observations on the trials of Gmelina arborea Roxb. at Gede and Jilori. Report prepared for Kenya Forest Dept. 15 pp.

Hughes, J. F., and R. M. A. Sardinha. 1975. The application of optical densitometry in the study of wood structure and properties. J. Microscopy 104(1):1-13.

Part 3

FERTILIZATION AND SITE GROWTH RESPONSE

15

FOLIAR ANALYSIS FOR PREDICTING QUANTITATIVE FERTILIZER RESPONSE: THE IMPORTANCE OF STAND AND SITE VARIABLES TO THE INTERPRETATION

R. Ballard and R. Lea

INTRODUCTION

Traditionally, nutrient deficiencies are diagnosed using nutrient concentrations in either plant tissue or soil. In forestry, foliar testing has generally proved superior to soil testing for identifying nutrient deficient areas (Ballard 1980). Considerable success has been achieved using foliar phosphorus (P) concentration as an index of response to P fertilization (Ballard and Pritchett 1975; Wells et al. 1973). However, total nitrogen (N) concentration in foliage has not proved particularly useful as an indicator of response to N fertilization (Mead and Gadgil 1978; Turner et al. 1977; Rosvall 1979). Some improvement in prediction of N status of forest stands has been reported using techniques such as N gradients over different age classes of tissue (Florence and Chuong 1974), N fractions in foliage (van den Driessche and Webber 1975), N-S interrelationships in foliage (Turner et al. 1977), and analysis of other biomass components such as litter (Miller and Miller 1976). While some of these techniques have been related to quantitative responses to N fertilization, none have been successfully related to quantitative responses over a range of site and stand conditions.

Recent efforts to explain variation in quantitative response of forest stands to N fertilization have tended to show that stand variables, in particular those providing some measure of stocking level, rather than nutritional variables are the best predictors (Duzan and Allen 1981; Mead and Gadgil 1978; Rosvall 1979; Strand and DeBell 1981). Such results appear reasonable where responses are recorded (or persist) for relatively short growth periods and the stands involved encompass a wide range of stand conditions but a restricted range of nutritional conditions. Intuitively we might expect a good prediction of quantitative response using a system that identifies sites according to both their nutrient status and their periodic annual increment over the response period--the nutrient status providing a measure of the _potential_ for growth improvement and the basic growth level providing a measure of the means by which quantitative expression is given to this potential.

This paper reports a study examining the relative value of stand, nutritional, and other site variables for predicting the quantitative response of loblolly pine (_Pinus taeda_ L.) to N and P fertilization.

METHODS

Thirty-one replicated trials were carefully selected from over one hundred NP fertilizer trials put into loblolly pine plantations by Forest Industry Cooperators of the North Carolina State Forest Fertilization

Cooperative. The trials were selected to provide a wide range of site and stand conditions and exhibited response to N and P fertilizers (Table 1). The trials are located throughout eight states in the southeastern United States and cover all major physiographic provinces in the region. Physiographic and ten-year average climatic properties were obtained for each site.

Soil and foliage samples were collected from the control plots of each trial. Soil samples were collected from the 0-10, 10-30, and 30-60 cm depths by compositing fifteen push tube samples for each plot. Profile pits were also dug to enable characterization of horizon depth, rooting depth, and drainage class (1-6, wet to dry). Composite foliage

Table 1. Stand and site characteristics of 31 loblolly pine plantations used in developing the fertilizer prediction model.

Variable	Mean Value	Standard Error of Mean	Range
Initial Stand			
Age (years)	8.6	0.9	3 - 22
Basal area (m^2/ha)	14	2.5	0 - 43
Volume (m^3/ha)	76	17	0.5 - 345
Response (5-year volume)			
To 112 kg N/ha (m^3/ha)	19	2.5	0 - 43
To 56 kg P/ha (m^3/ha)	12	2.2	0 - 61
Nutritional			
Foliar:			
Foliar N (%)	1.20	0.02	1.0 - 1.43
Foliar P (%)	0.11	0.002	0.09 - 0.14
Foliar K (%)	0.47	0.009	0.39 - 0.59
Foliar Mg (%)	0.12	0.0003	0.10 - 0.16
Foliar SO_4-S (ppm)	316	23	94 - 601
0-10 cm soil:			
Total N (ppm)	683	93	264 - 2,587
C/N ratio	31	1.1	19 - 43
KCl-N (ppm)	14	0.8	6 - 26
Anaerobic N (ppm)	19	1.8	3 - 46
Bray II P (ppm)	8.7	1.9	2.6 - 63
Other Soil (0-10 cm)			
Organic matter (%)	2.4	0.3	1.0 - 8.4
pH	4.7	0.09	3.8 - 6.3
Clay (%)	11	1.2	3 - 30
Site Index - 25 yr (m)	18	0.6	9 - 26
Depth A horizon (cm)	19	1.7	5 - 51
Rooting Depth (cm)	52.6	3.4	20 - >70
Drainage Class	4.4	0.2	2 - 6

samples from five trees were taken in January-March from the first flush of the previous season's growth on primary laterals in the upper one-third of the green crown.

Response variables were calculated as five-year gross volume response to N and P fertilization. The factorial design of the trials enabled the response to each element to be calculated after adjustment of base level growth for response to the other element. Nitrogen fertilizer was applied in early spring as ammonium nitrate at 112 kg N/ha. Phosphorus was applied at the same time as concentrated superphosphate at 56 kg P/ha.

Soil and foliage samples were analyzed by standard procedures for a range of properties and nutrient element concentrations. Tests were used to characterize levels of available N (total N, C/N, KCl extractable NH_4 + NO_3, aerobic and anaerobic mineralizable N) and available P (Bray I and II, Olsen, Double Acid, and New Mehlich). The selected foliar and soil properties are given in Table 1.

Data analysis involved an initial screening of the predictive capability of all variables using simple correlations. Model development was done in a systematic sequential manner examining groups of variables (stand, nutritional, and site) in turn. Within each group a variable was accepted into the model if it accounted for a significant amount of the variation in the residual variation of the model.

RESULTS AND DISCUSSION

Preliminary Screening

Preliminary screening of nutritional and site variables indicated that none of those measured were significantly correlated with volume response to N fertilization. Selected samples in Table 2 indicate the general poorness of the relationships. The poor predictive ability of both tissue and soil tests for N status is in agreement with the results of other workers (Mead and Gadgil 1978; Turner et al. 1977).

In contrast to N response, P response was significantly correlated with foliar P concentrations. Again, the superiority of tissue P levels is in agreement with other reports (Wells et al. 1973; Mead and Gadgil 1978). Volume response to P was not significantly correlated with any other nutritional and site variable tested (Table 2).

The only variables significantly correlated with volume response to N fertilization were those related to stocking levels such as age, volume, and basal area (Table 2). The quadratic form of the relationship provided a better fit than the linear form, indicating that beyond a certain stocking level response to N fertilizer will become less. Such a falloff presumably reflects a lack of space for crown expansion normally considered a prerequisite for response to N fertilization (Strand and DeBell 1981). The quadratic expression of basal area was also significantly correlated with the volume response to P fertilization (Table 2).

Nitrogen Response Model

The first step in the enhancement of the N prediction model based on the quadratic expression of basal area was to relate residuals from this model to the various nutritional variables. Foliar N was most closely

Table 2. Correlations (r^2) between response and soil/site/stand variables.

Variable	Response Variable	
	N	P
Nutritional		
Foliar N	0.098	
Foliar P		0.139*
Bray II P		0.001
Total soil N	0.056	
C/N ratio	0.934	
Site		
Rooting depth	0.003	0.037
Depth A horizons	0.009	0.009
Drainage class	0.035	0.098
Site index	0.048	0.031
Stand		
Age	0.207*	0.087
Volume	0.142*	0.021
Basal area (BA)	0.210*	0.070
BA + $(BA)^2$	0.279*	0.237*

*Significant at the 5% level.

correlated to the residuals and was entered into the model (Table 3). No other soil or foliar nutritional variables were significantly correlated with the residuals from the model including foliar N.

Residuals from the model including the basal area and foliar N variables were plotted against all site variables. Site index was positively and significantly correlated with residuals. Following entry of site index into the model, no other site variables or site x nutritional interactive variables were significantly correlated with model residuals.

The N response model with basal area, foliar N, and site index accounted for 43% of the variation in volume response to N fertilization and is illustrated graphically in Figure 1. While a model accounting for only 43% of the variation in the dependent variable may not be considered a powerful predictive tool, it is probably as good as can be expected for prediction of N response over a wide range of sites, since the effectiveness of N fertilizer is so strongly influenced by unpredictable climatic events around the time of application (Heilman et al. 1981). A further strength of such a model is its biological reasonableness: basal area provides a measure of the "capital" for building a response on; foliar N levels indicate the extent of the nutritional limitations and so opportunity for response through correction; and site index, with its nutritional component removed by virtue of its order in the regression, most likely reflects the moisture status control on

Table 3. Prediction models including (1) stand, (2) nutritional, and (3) site variables.

		N RESPONSE MODEL	R^2
1)	Y (m^3/ha/5 yr)	$= 9.2 + 1.21\ (BA) - 0.020\ (BA)^2$	0.279
2)	Y	$= 54 + 0.99\ (BA) - 0.014\ (BA)^2 - 37\ (FN)$	0.337
3)	Y	$= 51 + 1.01\ (BA) - 0.017\ (BA)^2 - 55\ (FN) + 1.38\ (SI)$	0.425
		P RESPONSE MODEL	
1)	Y (m^3/ha/5 yr)	$= 4.3 + 1.36\ (BA) - 0.030\ (BA)^2$	0.237
2)	Y	$= 97 + 1.14\ (BA) - 0.022\ (BA)^2 - 46\ (LFP)$	0.380
3)	Y	$= -551 + 1.36\ (BA) - 6.03\ (BA)^2 - 260\ (LFP) + 103\ (DRAIN) - 1.04\ (A\ DEPTH) + 49\ (LFP * DRAIN) + 0.3\ (A\ DEPTH * DRAIN)$	0.648

BA = basal area at fertilization (m^2/ha)
FN = foliar N concentration in unfertilized stand (%)
SI = site index, base age 25 years (m)
LFP = natural logarithm of foliar P concentration (%) in unfertilized stand
DRAIN = drainage class (1-6, using SCS classification)
A DEPTH = depth of A horizon (cm)

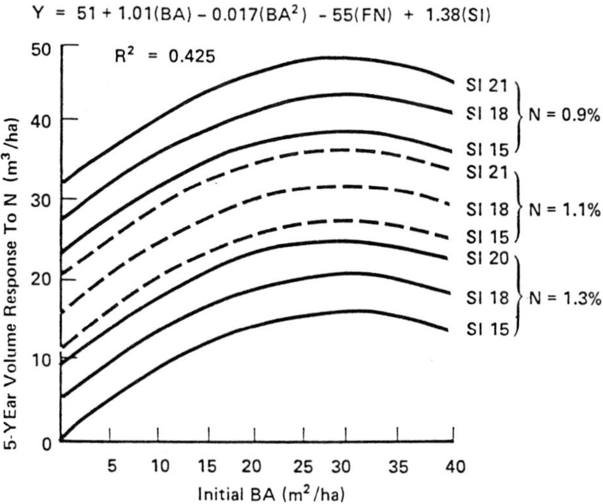

Figure 1. Relationship between N response and basal area, site index, and foliar N concentration at time of fertilization.

response expression, being the next most likely factor limiting to growth.

Phosphorus Response Model

The development and enhancement of the P response model followed the same steps used for the N response model. Residuals from the basic model involving basal area were plotted against and correlated with all nutritional variables. Foliar P was most closely related to model residuals and exhibited a curvilinear relationship. To accommodate the curvilinear form, foliar P was added to the model in the logarithmic form (Table 3). No other nutritional variables accounted for any significant variation in the residuals of the model following entry of log foliar P.

Examination of site and interactive variables showed that several of these variables made a significant contribution to reducing model variation. In order of entry into the model these variables were drainage class, depth of A horizon, and interactions Ln foliar P x drainage class and depth of A horizon x drainage class (Table 3).

The final P response model accounted for 65% of the variation in volume response to P fertilization. The relationship of basal area and foliar P to P response at fixed levels of drainage class and depth of A horizon is illustrated in Figure 2. The variables entering the model and the signs of the regression coefficients appear biologically sound, with higher volume responses over a five-year period occurring on well-stocked, P deficient, poorly drained sites with shallow A horizons.

The importance of the stocking level variable (basal area) in the model is undoubtedly determined in part by recording response over the limited time frame of five years. While such a limited time frame probably encompasses almost the entire period over which a response to N can be expected, this is not true for P fertilization where conventional application rates often provide responses that continue for decades (Ballard 1978). If P responses were evaluated over longer time frames,

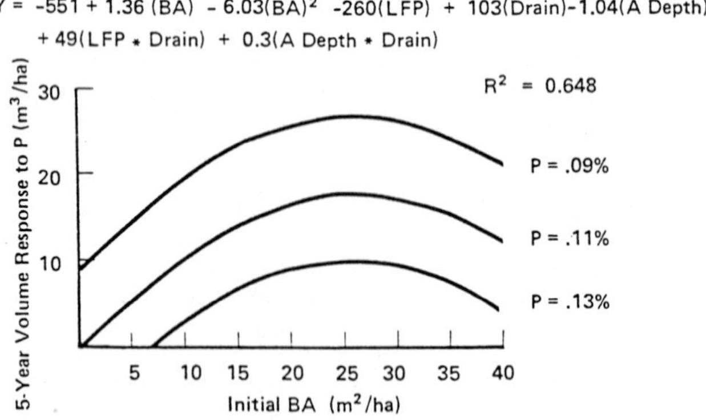

Figure 2. Relationship between P response and initial basal area for three levels of foliar P at mean values for drainage class (4.3) and depth of A horizon (19 cm).

it is reasonable to assume initial stand condition variables would assume less and nutritional variables more importance in the model.

CONCLUSIONS

Quantitative responses of forest plantations, assessed over relatively short periods of the rotation, are strongly influenced by the level of growing stock present at time of fertilization. The dominant effect of stocking level on response will most probably decline when assessing responses over major portions of the rotation. However, this will be true only for single applications of fertilizers providing long-term responses and multiple applications of fertilizers providing short-term resonses.

The dominant effect of stocking level has probably masked the relationship of nutritional variables to response in many calibration attempts where studies involving a range of initial stocking levels have been used. Such masking or confounding may lead to the erroneous conclusion that the technique used to assess the nutritional status has little value. In assessing and developing procedures for measuring nutritional status, it is imperative that the effect of stocking level on response be recognized. An understanding of the relative importance of these two determinants of fertilizer response for a variety of conditions is also most important in determining where a fertilizer investment will yield the best return.

REFERENCES

Ballard, R. 1978. Effect of first rotation phosphorus application on fertilizer requirements of second rotation radiata pine. N.Z. J. For. Sci. 8:135-145.

__________. 1980. The means to excellence through nutrient amendment. In: Forest plantations: The shape of the future, pp. 159-200. Weyerhaeuser Science Symposium 1. Weyerhaeuser Company, Tacoma, Washington.

Ballard, R., and W. L. Pritchett. 1975. Soil testing as a guide to phosphorus fertilization of young pine plantations in the Coastal Plain. Agr. Exp. Sta. Tech. Bull. 778. University of Florida, Gainesville. 22 pp.

Duzan, H. W., and H. L. Allen. 1981. Estimating fertilizer response in site prepared pine plantations using basal area and site index. In: J. P. Barnett, ed., Proc. First Biennial Southern Silvicultural Research Conference, pp. 219-222. USFS Gen. Tech. Rep. SO-35.

Florence, R. G., and P. H. Chuong. 1974. The influence of soil type on foliar nutrient concentrations in _Pinus radiata_ plantations. Aust. For. Res. 6:1-8.

Heilman, P. E., S. R. Webster, E. C. Steinbrenner, and R. F. Strand. 1981. Season for application of urea fertilizer to Pacific Northwest forests. In: Proc. Forest Fertilization Conference, pp. 186-191. University of Washington, Seattle.

Mead, D. J., and R. L. Gadgil. 1978. Fertilizer use in established radiata pine stands in New Zealand. N.Z. J. For. Sci. 8:105-134.

Miller, H. G., and J. D. Miller. 1976. Analysis of needle fall as a means of assessing nitrogen status in pine. Forestry 49:57-61.

Rosvall, O. 1979. Prognosfunktioner för berākning av gōdsling seffekter. Föreningen Skogsträdsforadling, Institutet för skogsforbattring, Årsbok.

Strand, R. F., and D. S. DeBell. 1981. Growth response to fertilization in relation to stocking levels of Douglas-fir. In: Proc. Forest Fertilization Conference, pp. 102-106. University of Washington, Seattle.

Turner, J., M. J. Lambert, and S. P. Gessel. 1977. Use of foliage sulphate concentrations to predict response to urea applications by Douglas-fir. Can. J. For. Res. 7:476-480.

van den Driessche, R., and J. E. Webber. 1975. Total and soluble nitrogen in Douglas-fir in relation to plant nitrogen status. Can. J. For. Res. 5:580-588.

Wells, C. G., D. M. Crutchfield, N. M. Berenyi, and C. B. Davey. 1973. Soil and foliar guidelines for phosphorus fertilization of loblolly pine. USDA For. Serv. Res. Pap. SE-110. 15 pp.

16

DIFFERENCES IN YIELD AT DIFFERENT SITES: AN IRRIGATION-FERTILIZATION STUDY OF NUTRIENT FLUX DURING FAST GROWTH

B. O. Axelsson

INTRODUCTION

Solar energy, water, and nutrients are necessary for plant growth, but nevertheless it has been difficult to construct a quantitative tree growth theory including the dynamics of water, nutrients, and carbon. This shows that we still do not understand the combined, dynamic effect of the essential elements that determine growth processes.

Several attempts to analyze tree growth have been made in the form of simulation models. These models either emphasize understanding of basic production processes or aim at predicting wood production per hectare and year. But models intended for prediction are seldom optimal for understanding, and vice versa. This apparent conflict points out the need for a deeper insight into the production processes--that is, a theory of tree growth that is as general as possible (Ågren 1981).

Many attempts to compare the productivity of the forests of the world have been severely limited by the small number of ecosystems that have been sufficiently documented to permit comparison. Therefore, the International Biological Program (IBP) studies of forests all over the world provided a unique opportunity. As a whole, however, the analysis revealed that forest productivity could not be described, and thus not understood, with the aid of simple relations like production as a function of temperature or precipitation. In fact, it is impossible to predict productivity from, for example, mean annual temperature or annual precipitation (cf. O'Neill and DeAngelis 1981).

Therefore, it might be argued that any analysis of differences in yield at different sites must have a more precise base. In the first place, productivity should be evaluated in relation to the nutrient flux through the system (Ingestad et al. 1981), because plant growth rate is proportional to the nutrient amount within the plant and this within a wide range of nutrient states (Ingestad 1982; Ågren 1983a, 1983b), and nutrient flux naturally varies considerably from site to site and is also fairly easy to manipulate. Historically, the introduction of foliar analysis as a method of assessing the internal nutrient levels of a plant offered a means of relating plant growth to the internal nutrient level instead of to the nutrient level in the substrate. As pointed out by Tamm (1968), for example, it is more or less impossible to interpret curves for growth response in relation to foliar concentrations obtained from field experiments, since the responses obtained for growth and foliar levels are not simultaneous. Thus there may be any type of relationship between growth and nutrient concentrations (cf. Ingestad 1982; Axelsson 1982). Instead of working with strongly variable nutrient concentrations within the plants, we must maintain a more or less constant level in the foliage over a period of time, long enough to be able to measure a reliable growth response (Tamm 1968). In addition, the root/shoot ratio is a function of nutrient flux, which means that the nutrient flow rate through the system not only determines the productivity level but also the partitioning between organs (Ingestad and Lund 1979).

In this study it is hypothesized that the nutrient flux during the period of fast growth is the key variable for understanding and predicting forest productivity. A theory for forest productivity is applied to an irrigation-fertilization field experiment with Scots pine in mid-Sweden. The analysis demonstrates the dynamic effects of fertilization on the fertility level of the ecosystem--that is, it offers guidelines for fertilization in practice.

FOREST PRODUCTIVITY AND THE NUTRIENT FLUX

Irradiation is a strong driving force for plant productivity, but a change in leaf production with a factor 10 to 15 corresponds to a change in absorbed energy on the earth's surface of only a factor 2 (cf. O'Neill and DeAngelis 1981; Liljequist 1970).

Furthermore, existing data indicate that rates of carbon influx to conifer canopies compare well with estimates of carbon influx for agricultural crops under the same climatic conditions (Jarvis 1981). Indeed, published values of photosynthetic capacity differ so little among land plants in general, with the exception of C_4- plants, that there is no final evidence of different productivity as a result of different photosynthetic capacity.

There are good reasons to suggest that nutrient status, especially nitrogen level, within the plant is the determinant of plant growth rate. Consequently, almost any natural forest is far from maximum productivity under prevailing climatic conditions because of suboptimal nutrient circumstances. Therefore, it is not surprising that very high dry-matter production figures are reported from irrigation-fertilization experiments for numerous different woody species.

The basic nutrition-growth relationships are reviewed by Ingestad (1982), but two important aspects should be stressed here. The relative growth rate for any plant species is a linear function of nutrient flux, but maximum growth rate for a species (b) is lower than for a species (a) because of some environmental stress factor (Figure 1A). It is also important to note that the same nutrient concentration within the two species means different growth rates (Figure 1B). As mentioned above, it is more or less impossible to interpret growth responses in relation to foliar concentration data, not only because of unstable nutrient conditions but also because any varying stress factor might influence the inclination between the relative growth rate and nutrient concentration.

SCOTS PINE PRODUCTIVITY: IRRIGATION-FERTILIZATION EXPERIMENT

Mainly to test these ideas about productivity, a field experiment with a fifteen-year-old Scots pine stand in mid-Sweden was started in 1974 (Aronsson, Elowson, and Ingestad 1977). The experiment shows the differences in ecosystem response between irrigation, annual applications of solid fertilizer in single large doses, and daily liquid applications during the growing seasons. The experimental layout was five randomized blocks, and in the autumn of 1979 twenty representative trees, chosen on

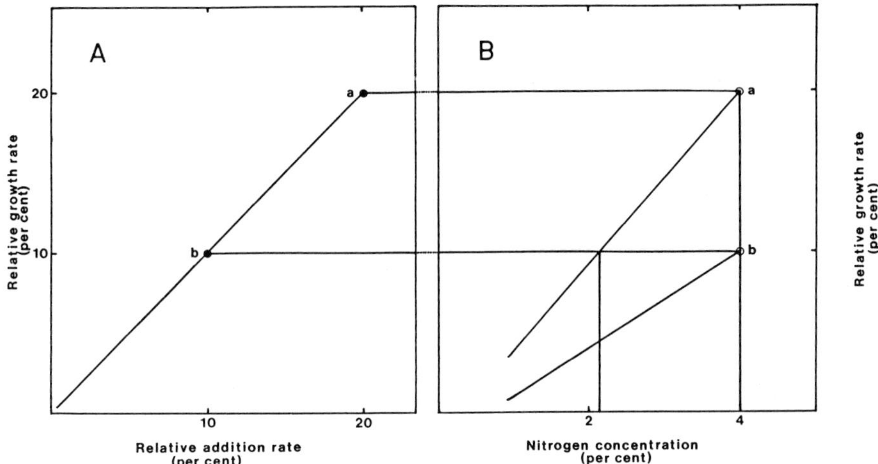

Figure 1. Relationship between A (relative addition rate of nitrogen and relative growth rate) and B (nitrogen status of the plant and relative growth rate) for two plant species or two hypothetical environments (generalized, after Ingestad 1981).

the basis of height and diameter distribution, were analyzed with regard to biomass and net production above and below ground. In the following autumn, 1980, a total harvest of all trees on one block was done, after which the basal areas of the remaining plots reflected the overall effects after 7.5 treatment seasons (inventory 1981), Figure 2. The basal areas indicate a possible but small irrigation effect and also a small positive effect of liquid fertilization compared with solid fertilizer application. An analysis of the covariance for all five blocks in 1979 revealed that the effect of fertilization was highly significant ($p < 0.999$). Irrigation also resulted in significant response ($p < 0.99$), while there was no significant interaction effect. These results are mainly confirmed by an analysis of the relative increments of the basal areas for the whole experimental plots (Figure 3) and for harvested representative trees from block no. 1--that is, plots 1-4 (Figure 4), respectively.

It is necessary to have these general outcomes in mind when looking at the biomass figures from the 1979 sampling (Table 1). Thus, an annual growth rate of the order of 8% as an effect of irrigation only (Table 1) is confirmed by (and also corresponds to) an accumulated difference in stem volume of about 70% over a seven-year treatment period (Table 2). It is notable, however, that all treatments mean a change in the allocation pattern of carbon, with relatively more of the net biomass above ground and less below ground (Table 3). Irrigation seems to reduce the relative coarse root production (Table 3), while accumulation of fine root biomass increases (Figure 5). Excavation of whole tree coarse root systems supports these results, demonstrating that irrigated trees had thinner main roots close to the stump but much more secondary ramifications. On the irrigated plots these roots were mainly horizontal, while plots treated with liquid fertilizer had trees with more vertical ramifications penetrating deeper into the soil profile, probably a reflection of the soil fertility development.

Table 1. Annual biomass (dw) production, 1979, in a 20-year-old Scots pine stand at Ivantjärnsheden (kg/ha/yr). Relative treatment effects in parentheses (%) (recalculated from Axelsson 1981; estimates of fine root production from Persson 1978).

Treat-ment*	Needles	Branches cur-rent	Branches old	Stems	Stumps	Roots coarse (>2mm)	Roots fine (<2mm)	Total
0	1,160 (100)	320 (100)	680 (100)	780 (100)	140 (100)	810 (100)	1,830 (100)	5,700 (100)
I	1,290 (111)	400 (126)	860 (126)	1,060 (135)	190 (139)	580 (71)	1,830 (100)	6,200 (108)
F	2,740 (236)	960 (304)	1,560 (228)	1,920 (245)	340 (247)	2,420 (301)	1,830 (100)	11,800 (206)
IF	3,210 (276)	1,040 (330)	1,740 (254)	2,290 (292)	460 (334)	1,820 (226)	1,830 (100)	12,400 (217)

*0 = control, I = daily irrigation, F = solid fertilization once a year, IF = daily liquid fertilization. Treatment period: late summer 1974-79.

Table 2. Stem volumes in 1981 (m^3/ha) and average annual gain over control, 1975-81 (m^3/ha/yr), in stand II, blocks 3-5, Ivantjärnsheden.

Block No.	Control	Irrigation	Solid fertilization	Liquid fertilization	Mean
	Stem volumes				
3	10.7	18.8	29.3	37.8	24.1
4	16.2	21.9	34.4	35.5	27.0
5	8.9	19.8	38.5	44.1	27.8
Mean (s.e.)	11.9 (2.2)	20.2 (0.9)	34.1 (2.7)	39.1 (2.6)	26.3
	Annual gain				
Average annual gain	-	1.2	3.2	3.9	
Annual growth rate relative to control	1.0	1.08	1.16	1.19	

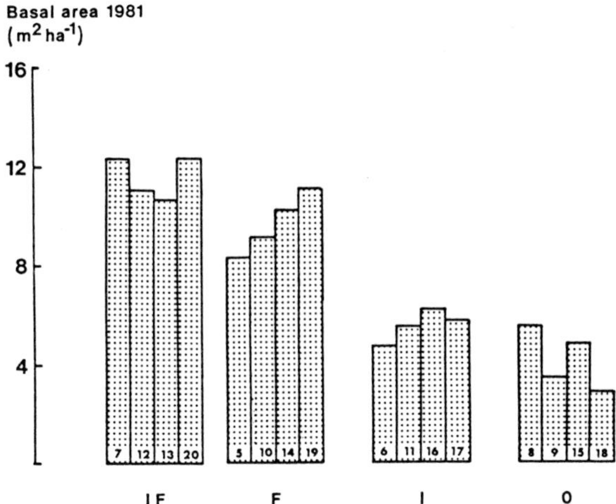

Figure 2. An irrigation and fertilization experiment was started in July 1974 at Ivantjärnsheden, Sweden. After 7.5 treatment seasons, the differences in stem basal areas of the approximately twenty-year-old Scots pine stands were significant as an effect not only of fertilization but also of irrigation. Treatments: IF = daily liquid fertilization, F = annual solid fertilization, I = daily irrigation, and 0 = control.

Table 3. Relative distribution of annual biomass (dw) production (%), 1979, in a 20-year-old Scots pine stand at Ivantjärnsheden (recalculated from Axelsson 1981).

		Branches				Roots		
Treat-ment*	Needles	cur-rent	old	Stems	Stumps	coarse (>2mm)	fine (<2mm)	Total
0	20.3	5.5	12.0	13.7	2.4	14.1	32.0	100.0
I	20.7	6.4	13.9	17.0	3.1	9.3	29.5	99.9
F	23.3	8.1	13.3	16.3	2.9	20.6	15.5	100.0
IF	25.9	8.4	14.0	18.5	3.7	13.7	14.8	100.0

*See Table 1.

In conclusion, irrigation leads to increased productivity of aboveground biomass but decreased productivity of roots, and fertilization (in particular liquid applications) means a pronounced shift of production toward aboveground organs (Table 4).

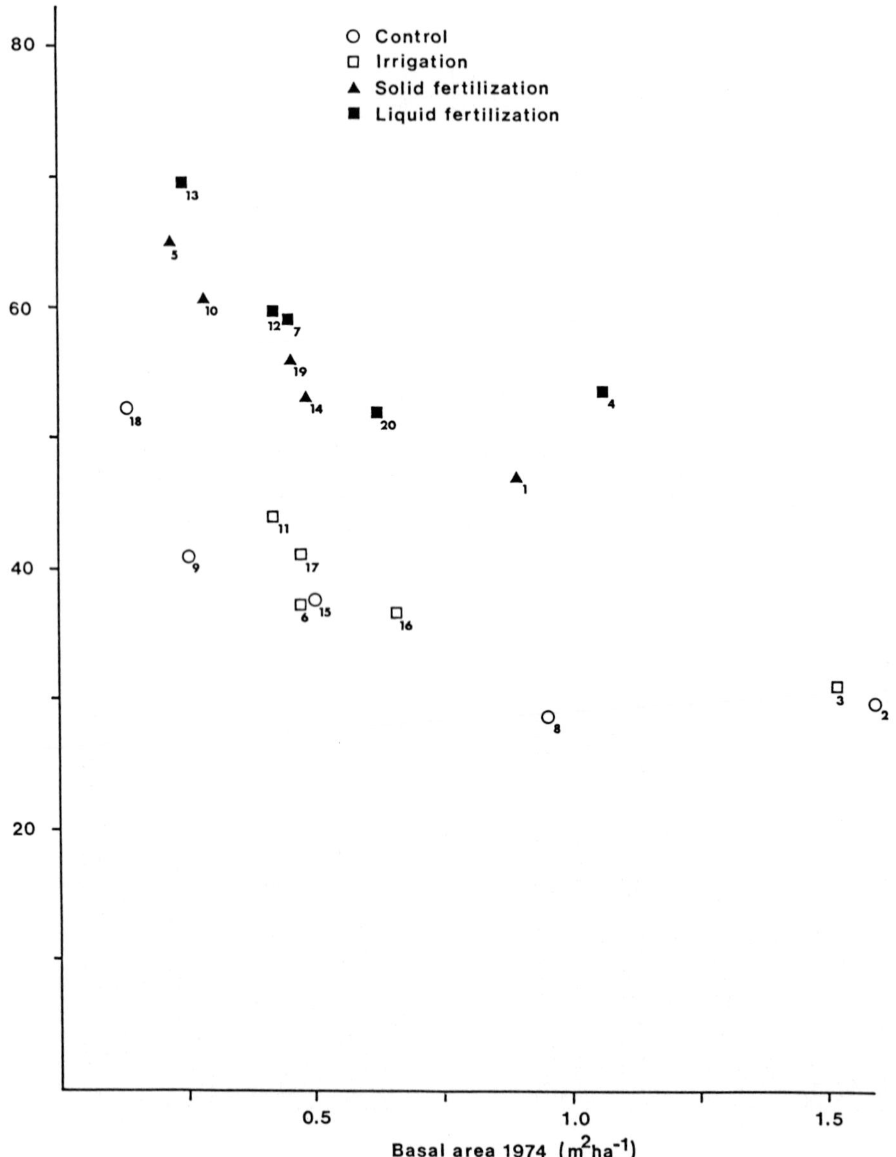

Figure 3. Relative annual growth rate per experimental plot, 1974-79, as a function of basal areas at the beginning of the experiment at Ivantjärnsheden. Plot numbers are the same as in Figure 2.

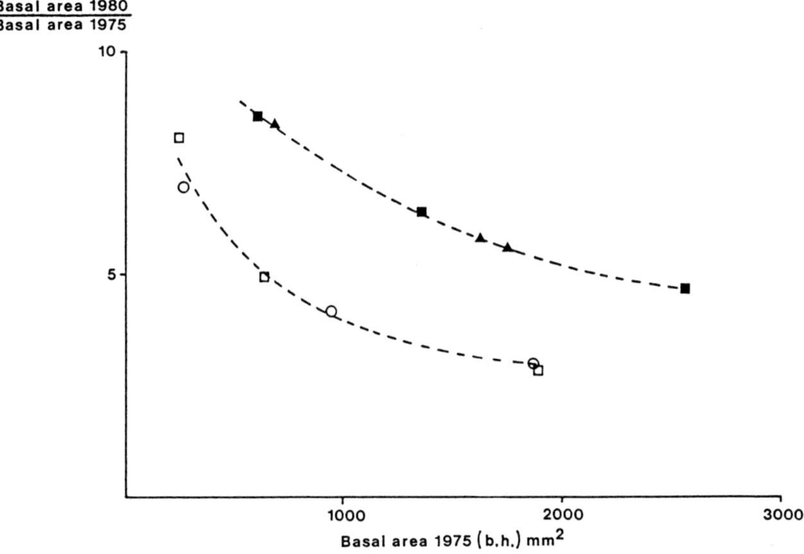

Figure 4. Relative annual growth rate of the basal area, 1975–80, as a function of tree size, for a sample of representative trees from the irrigation and fertilization experiment at Ivantjärnsheden. Symbols are the same as in Figure 3.

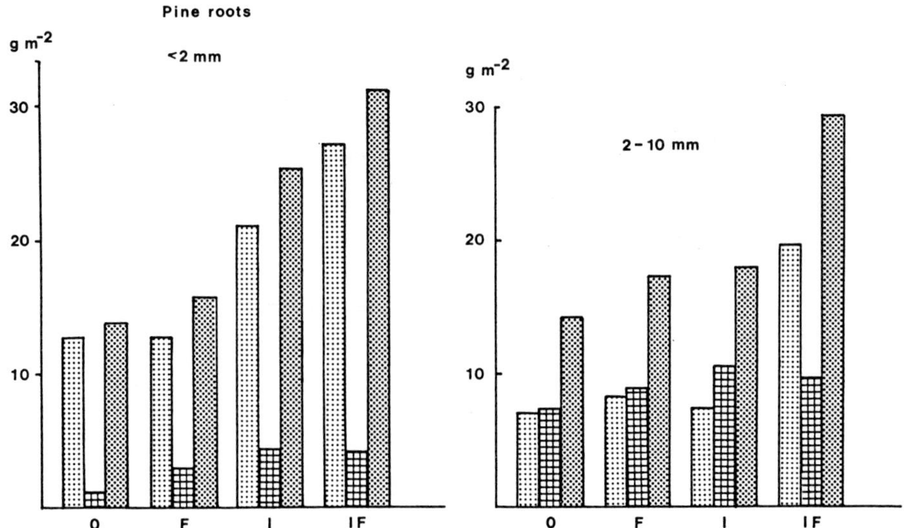

Figure 5. The effect of irrigation and/or fertilization on root biomass (left bars), root necromass (central bars), and biomass plus necromass (right bars). The sampling was carried out in October 1979 at Ivantjärnsheden. By the courtesy of H. Persson (unpubl.).

RELATION BETWEEN THEORY AND REALITY

Fertilization experiments, like the one referred to above, mainly aim at understanding nutrition-growth relations but also give guidelines for practical applications. Both aspects are summarized in Table 5. As expected, fertilization implied increased production but reduced relative productivity per unit of leaf mass because of self-shading (cf. also Table 4). For a more detailed evaluation of treatment effects reference must be made to the relationships in Figure 1. In theory and under optimal conditions a Scots pine stand will reach the maximum growth rate (a), (disregard the parameter values in Figure 1). The interference of any growth-limiting variable, except nitrogen, reduces the relative growth rate for a given nitrogen status down to, for example, the line with maximum rate (b). This means that great care must be taken in the interpretation of nutrient concentration data from field fertilization experiments. The nitrogen content in exposed current needles sampled in September each year during 1974-80 indicates initial treatment effects which should lead to a higher growth rate as a consequence of irrigation, but also to higher growth rate on plots treated with solid fertilizer than those with liquid applications. Over time, however, the nitrogen status is changed. Irrigated plots have established equilibrium below the control plots, and the difference between the two fertilization alternatives is leveling out (Figure 6). As shown above, the irrigated plots were growing at a relative rate significantly larger than that in the control plots, whereas there was no significant difference between the fertilization regimes until 1979 as regards stem biomass production. It should be noted, however, that the lower nitrogen status on the irrigated plots than on the control plots corresponds to the same amount of nitrogen on an area basis and gives the same total biomass production per unit of needle mass (Table 5). Thus, irrigation initially means that the mineralization is stimulated and the amounts of nitrogen in the foliage are increased with increased foliage production. This stimulating effect can, however, not be maintained, and the net uptake in the irrigated plots decreases relative to control plots, eventually giving equal amounts of nitrogen in the foliage. As a whole, the turnover of nitrogen through the system is controlling productivity. The treatment effect is in principle only a gain in time, and there is no

Table 4. Productivity (kg dw/kg dw of needles) in a 20-year-old stand of Scots pine (based on Axelsson 1981; Linder and Axelsson 1982).

Treatment	Shoots	Branch Wood	Stems	Stumps	Roots	Net Photo-synthesis	Respi-ration
Control	0.47	0.22	0.25	0.04	0.83	2.4-2.7	0.6-0.8
Irrigation	0.48	0.25	0.30	0.06	0.69	2.4-2.7	0.6-0.8
Solid fer-tilization	0.51	0.22	0.27	0.05	0.59	2.1-2.4	0.5-0.7
Liquid fer-tilization	0.57	0.23	0.31	0.06	0.49	2.1-2.4	0.5-0.7

Table 5. Biomass (kg dw/ha) and relative productivity, 1979, in a 20-year-old Scots pine stand, Ivantjärnsheden.

			Total Annual Production	
Treatment*	Needle biomass	Stem biomass	Needle biomass	Nitrogen supply
Control	3,200 (100%)	4,100 (100%)	1.81 (100%)	--
Irrigation	3,500 (110%)	4,900 (120%)	1.78 (98%)	--
Solid fertilization	7,200 (228%)	8,400 (205%)	1.64 (90%)	25
Liquid fertilization	7,400 (235%)	9,400 (228%)	1.67 (92%)	14

* Treatments: Control, daily irrigation during growing season, annual solid fertilization (total application until 1979 equal to 480 kg N, 120 kg K, and 60 kg P per hectare), and daily liquid fertilization during growing season (total application until 1979 equal to 870 kg N per hectare, all other essential nutrients in proportion to nitrogen).

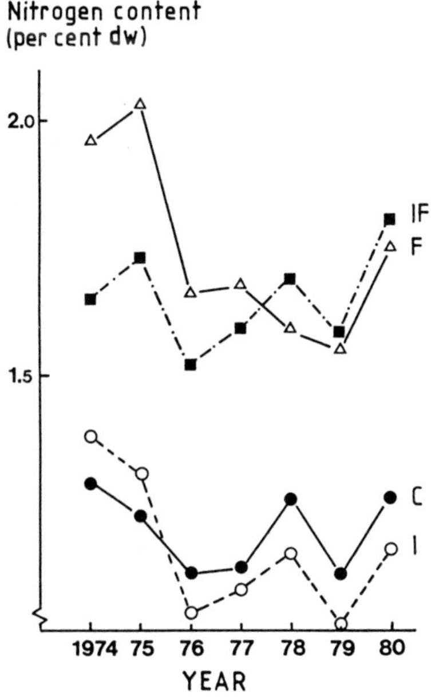

Figure 6. Nitrogen content as percentages of dry weight in exposed needles taken from the third whorl from the top of the trees. Each value is an average of five plots at Ivantjärnsheden, and all sampling was done in September 1974-80.

change in productivity. Because of a change in allocation of carbon, there is, so far, a gain in stemwood production (Tables 1-5).

In practice, fertilization today--that is, application of one or a few nutrient elements--always means a risk of disturbed proportions between nutrient elements and thus disturbed growth. In the first place, conventional application rates are far above the current consumption capacity of the crop systems treated. Then, with regard to theory, it is relevant to discuss the effects of growth-limiting factors and injurious effects. Both types of effects might be responsible for reduced relative growth rates according to Figure 1B. In the first place, there is good reason to discuss growth-limiting factors, among which water stress might be a significant consequence of fertilization. Also in field experiments aiming at balanced nutritional applications matching nutrient turnover without losses from the system, deviations from optimum nutritional proportions might come in. The significance of keeping the nutrient status in balance is clearly demonstrated by an optimum nutrition experiment with Norway spruce, where application with nitrogen alone will reduce growth rate to half that of plots given complete fertilizer treatment (Figure 7). Also in the Scots pine experiment at Ivantjärnsheden, height growth was disturbed in the third season after start and became more pronounced during the next two years as a result of fertilization irrespective of application technique. As there were no micronutrients in the solid fertilizer, it was not surprising that typical boron deficiency symptoms appeared on these plots. But why did the same result appear on the plots given liquid fertilizer where the applications included all nutrients in balanced proportions? Finally, the cause was traced back to the nutrient solution used (bought commercially), which was found to deviate too much from recommended proportions (Ingestad 1979). Thus accidentally, the addition rate of boron became only about half of the calculated need, in turn causing a significant disturbance of leading shoot growth. Leading shoot length was significantly longer on fertilized plots during 1975-78, but in 1979 there was no difference (Flower-Ellis 1982).

CONCLUSION

Variation in forest productivity from site to site is foremost a function of the nutrient flux through the system. Nitrogen is always the key element from an physiological point of view, and, in addition, nitrogen is most likely limiting for a majority of forest ecosystems throughout the world. Over large areas, nitrogen alone is growth limiting, while in other systems with intermediate to good nitrogen status, the interaction with other nutrient elements or other growth factors has reduced productivity to a level far from the biological maximum.

There is much evidence suggesting that forest productivity can be notably increased by fertilization, but the application technique in future practice must be adjusted to the biological capacity of the system treated in order to avoid ecologically unacceptable consequences and to maximize efficiency.

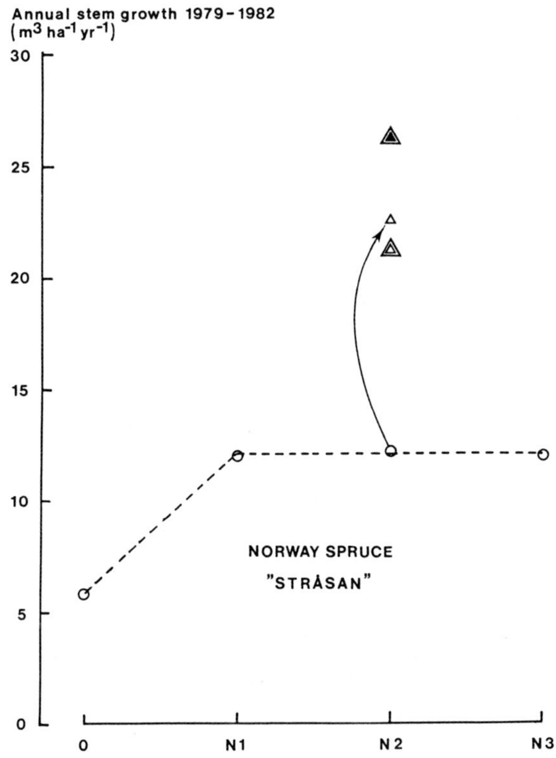

Figure 7. Annual stem growth of Norway spruce, 1979-82, as affected by different fertilization regimes. The treatments N1-N3 are multiple application rates of nitrogen (open circles). Fertilization with all essential macro- and micronutrients means a double growth rate (open triangles) over the nitrogen response. The experiment at Stråsan also includes plots with application of lime (large filled triangle).

ACKNOWLEDGMENTS

Dr. G. I. Ågren has contributed a number of helpful comments, for which the author is most grateful. The investigation was supported by grants from the Swedish Council for Forestry and Agricultural Research.

REFERENCES

Ågren, G. I. 1981. Problems involved in modelling tree growth. In: S. Linder, ed., Understanding and predicting tree growth. Studia for. suec. 160:7-18.

__________. 1983a. Nitrogen productivity of some conifers. Can. J. For. Res. 13(3):494-500.

__________. 1983b. The concept of nitrogen productivity in forest growth modelling. Mitteilungen der Forstlichen Bundesversuchanstalt 147:199-210.

Aronsson, A. 1984. Förändring i trädens mineralsammansättning vid gödsling. Skogsgödsling och miljön, 1984.02.10. In press.

Aronsson, A., and S. Elowson. 1980. Effects of irrigation and fertilization on mineral nutrients in Scots pine needles. In: T. Persson, ed., Structure and function of northern coniferous forests: An ecosystem study. Ecol. Bull. (Stockholm) 32:219-228.

Aronsson, A., S. Elowson, and T. Ingestad. 1977. Elimination of water and mineral nutrition as limiting factors in a young Scots pine stand. I. Experimental design and some preliminary results. Swed. Con. For. Proj. Tech. Rep. 10. Uppsala. 38 pp.

Axelsson, B. 1981. Site differences in yield: Differences in biological production or in redistribution of carbon within trees. Swed. Univ. Agr. Sci. Dept. Ecol. Environ. Res. Rep. 9. Uppsala. 11 pp.

__________. 1982. Potential production in Scots pine: Principles of plant productivity. Skogs- o. Lantbr.-akad. Tidskr. Suppl. 14:51-57.

Flower-Ellis, J. 1982. Structure and growth of some young Scots pine stands: 2. Effects of irrigation and fertilization on the amount, rate and duration of leading shoot growth. Swed. Con. For. Proj. Tech. Rep. 30. Uppsala. 46 pp.

Ingestad, T. 1979. Mineral nutrient requirements of *Pinus sylvestris* and *Picea abies* seedlings. Physiol. Plant. 45:373-380.

__________. 1981. Nutrition and growth of birch and grey alder seedlings in low conductivity solutions and at varied relative rates of nutrient addition. Physiol. Plant. 52:454-466.

__________. 1982. Relative addition rate and external concentration: Driving variable used in plant nutrition research. Plant, Cell and Environment, pp. 443-453.

Ingestad, T., A. Aronsson, and G. I. Ågren. 1981. Nutrient flux density model of mineral nutrition in conifer ecosystems. Studia for. suec. 160:61-71.

Ingestad, T., and A. B. Lund. 1979. Nitrogen stress in birch seedlings. I. Growth technique and growth. Physiol. Plant. 45:137-148.

Jarvis, P. G. 1981. Production efficiency of coniferous forest in UK. In: C. B. Johnson, ed., Physiological processes limiting plant productivity, pp. 81-107. The Camelot Press, Southampton.

Liljequist, G. H. 1970. Klimatologi. Generalstabens litogr. anstalts förlag, Stockholm.

Linder, S., and B. Axelsson. 1982. Changes in carbon uptake and allocation patterns as a result of irrigation and fertilization in a young _Pinus sylvestris_ stand. In: R. H. Waring, ed., Carbon uptake and allocation in subalpine ecosystems as a key to management, pp. 38-44. IUFRO Workshop, Corvallis, Oregon.

O'Neill, R. V., and D. L. DeAngelis. 1981. Comparative productivity and biomass relations of forest ecosystems. In: D. E. Reichle, ed., Dynamic properties of forest ecosystems, pp. 411-449. Cambridge University Press, Cambridge.

Persson, H. 1978. Root dynamics in a young Scots pine stand in Central Sweden. Oikos 30:508-519.

Tamm, C. O. 1968. An attempt to assess the optimum nitrogen level in Norway spruce under field conditions. Studia for. suec. 61:1-67.

17

THE RESPONSE OF SOME NORWAY SPRUCE (*PICEA ABIES* [L.] KARST.) PROVENANCES TO FERTILIZERS

N. Komlenović and J. Gračan

INTRODUCTION

The establishment of forest cultures is one of the most complex of forest activities. The choice of the proper seed origin is of special importance. But because of an insufficient number of provenance experiments, little is known about the growth of our domestic forest species outside their natural range. That applies also to the Norway spruce, one of the species that has been planted the most in Yugoslavia. Little is known about the differences in growth between its provenances or the appropriate sites to establish new cultures and application of mineral fertilizers.

Some of those who have reported about the interspecific differences of forest tree species with regard to fertilizers are Fielding and Brown (1961), Kral (1961), Ingestad (1963), Walker and Hatcher (1965), Giertych (1968), Squillace (1969), Zobel and Roberds (1970), Evers (1980), Chopin et al. (1982), and Komlenović and Krstinić (1982).

EXPERIMENTAL METHODS

In the typical sites of Norway spruce forests in Croatia (southwest Yugoslavia), fifteen representative stands were chosen. During a three-year period, samples of needles were collected and analyzed. After these investigations, in ten stands of Norway spruce the seed was collected. In that sampling we included one provenance of both Omorika (*Picea omorika* Panč.) and Sitka spruce (*Picea Sitchensis* Carr.). The experiment was laid out in the nursery (130 m altitude). The split-plot design with five replications was applied. The treated plots received NPK fertilizers.

RESULTS

All analyzed Norway spruce stands within its natural range differed considerably with regard to growth. The heights of dominant trees in the hundredth year ranged from 10.0 to 35.2 m. Despite different pedological conditions, a strong negative correlation has been found between the mentioned heights and the corresponding altitude (r = -0.53). The multiple regression analysis showed that 81% of variation in the height growth can be explained by nitogren, phosphorus, potassium, magnesium, and calcium concentrations in the needles. The strongest correlation was established between the growth and concentrations of nitrogen. The concentrations of nitrogen in one-year-old needles varied from 1.22% to 1.44%.

Young Norway spruce cultures established below the natural range on pseudogley soils grow slowly and have yellow needles with a nitrogen concentration often below 1%. With the application of nitrogen fertilizers in these cultures, the positive results in growth are regularly achieved. The concentration of phosphorus in one-year-old needles is in most cases over 0.25%, and the application of this nutrient does not give positive results.

In cultures established on distric cambisol on terra rossa, tera rosa luvic, and luvisol acric under the vegetation of *Calluna vulgaris* and *Pteridium aquilinum*, we have observed the lack of nitrogen, but also the lack of phosphorus. The concentration of phosphorus in one-year-old needles is often less than 0.1%. The application of nitrogen gives usually small effects. But after the phosphorus application and the increase of its content in the plant, the effect of nitrogen becomes significant.

Table 1 provides data on the average values of dry-matter content. On the basis of these data it has been proved that the applied fertilizers had a positive effect on the growth of the plants of all analyzed provenances. The interaction of provenance x fertilizers has not been found with regard to the growth of two-year-old seedlings.

The dry-matter content of Omorika and Sitka spruce was considerably less than of Norway spruce. The differences between some of the analyzed provenances have been found. Results have shown a negative correlation between the dry-matter content from two-year-old seedlings and corresponding altitude of the tested provenances. On the NPK plots the linear correlation was $r = -0.59$, but on the unfertilized plots it was $r = -0.34$. If we count only the dry-matter content of the needles and stem, the correlation stronger, because the corresponding values are $r = -0.77$ and $r = -0.62$. The proclaimed regularity is understandable.

The quantities of analyzed nutrients in two-year-old seedlings (Figure 1) showed a positive and strong correlation with plant dry-matter content.

Results have shown that in establishing intensive forest cultures below the natural range of Norway spruce, special attention should be paid to the provenance from the lower altitude. They normally grow faster, and, with the use of mineral fertilizers, better results can be expected.

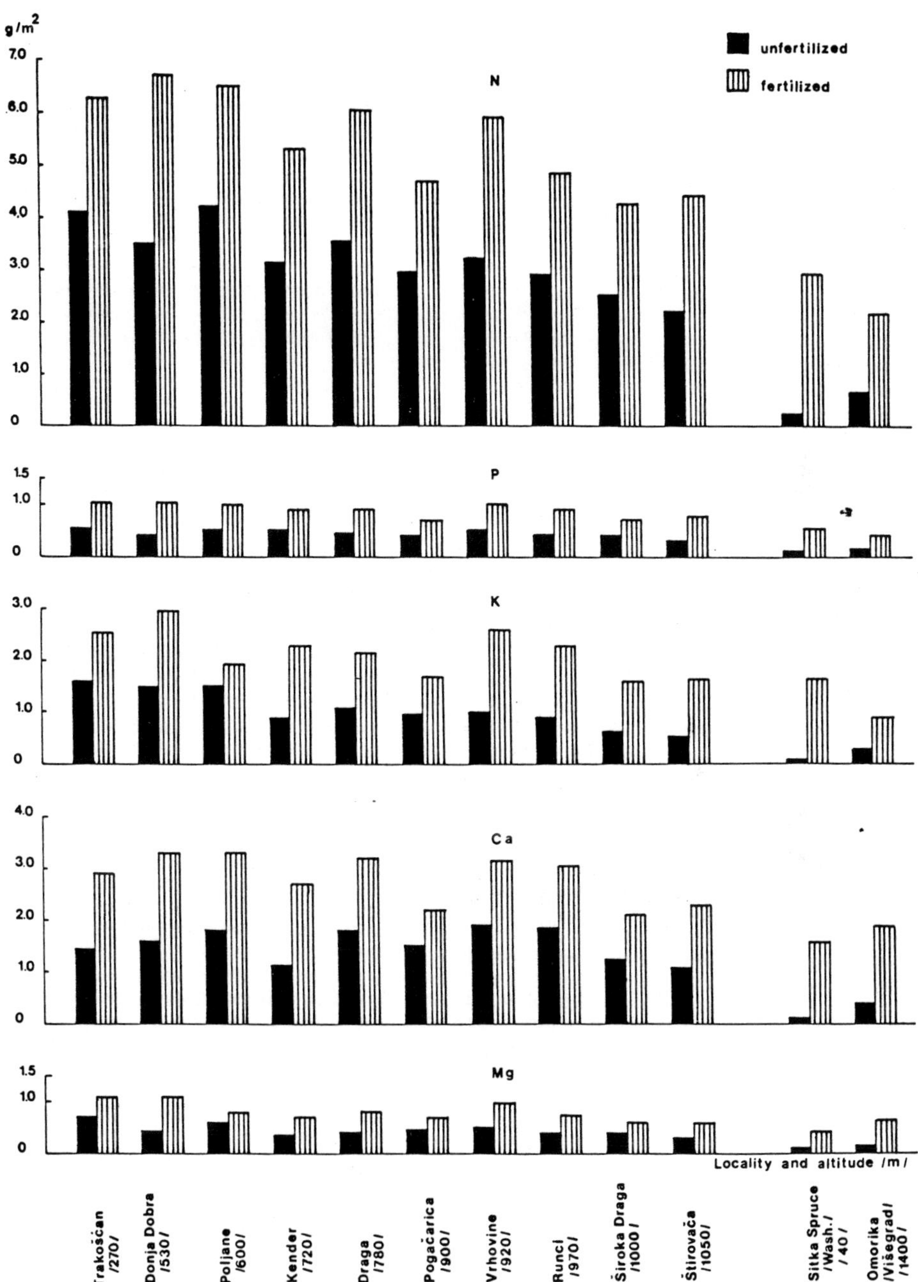

Figure 1. The quantities of nutrients in the aboveground part of two-year-old seedlings according to altitude.

Table 1. Average values of dry-matter content (g/50 plants).

Treatments	Variants		Provenance Mean
	NPK	Unfertilized	
So-1161, Poljane	60.62	34.80	47.71
So-1163, Draga	53.98	29.81	41.90
So-1259, Štirovača	46.21	25.15	35.68
So-1260, Vrhovine	63.46	30.88	47.17
So-1261, Široka Draga	41.75	21.66	31.71
So-1263, Runci	59.55	25.84	42.63
So-1264, Pogačarica	46.88	25.95	36.21
So-1265, Kender	51.63	27.44	39.53
So-1266, Donja Dobra	68.06	30.13	49.10
So-1268, Trakošćan	61.05	32.07	46.56
So-1270, Sitka spruce	30.63	2.32	16.47
So-1271, Omorika spruce	22.65	5.17	13.91
Treatment mean	50.50	24.27	37.39

"t"$_{05}$ = 5.67 for provenances "t"$_{05}$ = 14.11 for fertilization

"t"$_{01}$ = 7.57 "t"$_{01}$ = 19.91

REFERENCES

Chopin, F. S. III, P. K. Tryon, and K. Van Cleve. 1982. Influence of phosphorus supply on the growth and biomass allocation of Alaskan taiga tree seedlings. Institute of Arctic Biology, Fairbanks. 25 pp.

Evers, F. H. 1980. Untersuchungen an zwei Fichten Klonen bei unterschiedlicher Ernährung: Ein Beitrag zur nadelanalytischen Metodik. Sonderdruck aus Der Forst- und Holzwirt. 3 pp.

Fielding, J. M., and A. G. Brown. 1961. Tree-to-tree variations in the health and some effects of superphosphate on the growth and development of Monterey pine on a low-quality site. Commonwealth of Australia, For. and Timber Bur. Leaflet No. 79. 19 pp.

Giertych, M. M. 1968. Variations in Norway spruce provenances in their ability to utilize mineral nutrients available in limited quantities under competitive conditions. Fourth Annual Rep., Polish Acad. Sci., Inst. of Dendrology and Kornik Arboretum. 38 pp.

Ingestad, T. 1963. Comparison of nutritional properties in forest tree species. For. Cen. and Tree Impr., Stockholm, Sec. 5/9. 5 pp.

Komlenović, N., and A. Krstinić. 1982. Genotipske razlike nekih klonova stablastih vrba s obzirom na stanje ishrane. Topola 26:29-39.

Kral, E. 1961. Untersuchungen über den Nährstoffhaushalt von auf gleichem Standort erwachsenen Fichtenjungpflanzen in Abhängigkeit von ihrer Wuchsenergie und Herkunft. Cbl. Ges. Forstwes. 78:18-38.

Squillace, A. E. 1969. Field experiences on the kinds and sizes of genotype-environment interaction. IUFRO Working Group on Quantitative Genetics, Section 22, Raleigh.

Walker, L. C., and R. D. Hatcher. 1965. Variation in the ability of slash pine progeny groups to absorb nutrients. Soil Sci. Soc. Am. Proc. 29(5):616-621.

Zobel, B., and J. Roberds. 1970. Differential genetic response to fertilizers within tree species. Society of American Foresters, Michigan. 19 pp.

THE FERTILIZER TRIAL OERREL IN *PICEA ABIES*, WITH REFERENCE TO SAMPLING METHODOLOGY

A. Van Laar and J. B. Reemtsma

INTRODUCTION

In 1930 a fertilizer trial was established in the Forest District Oerrel on the Luneburger Heide, in North Germany, to assess the effect of fertilization on growth of *Picea abies*. The soil is a moderately poor to poor podzolized sandy soil and is of glacial origin. The experimental area is 90 m above sea level. The mean annual rainfall is 750 mm, of which 340 mm occurs during the growing season. Mean annual temperature is 8.3°C and the mean temperature during the growing season is 14.9°C.

The experiment represents the classical deficiency trial and was established with seven fertilizer applications and three replications in an unrestricted random layout. Experiment treatments tested and quantities of fertilizer used are shown in Tables 1 and 2.

Table 1. Experiment treatments.

Treatment	Fertilization 1930-58	Fertilization in 1970
1	Control	
2	NPKCa	NK
3	LPKCa*	-
4	PKCa	K
5	NKCa	NK
6	NPCa	N
7	NPK	NK

*L = Lupin.

Table 2. Quantities of fertilizer used in tests (kg/ha).

Years	N	P_2O_5	K_2O	CaO
1930-49	117	100	129	1,540
1956-58	100	96	42	1,160
1970	100	-	90	-

In 1976, five trees were selected at random from the stratum of healthy dominants in each of the twenty-one plots. They were sampled by destructive sampling. The present study deals with foliar analysis and growth of these sample trees.

Needle samples were taken from the main axis of the primary branch and from side branches on both sides of the primary branch. They were obtained from two positions, representing the top and seventh whorls respectively. The seventh whorl was represented by the needle ages between one and six years. The relevant positions and ages are described as 1/1, 7/1, 7/2...7/6. All branches from the top and seventh whorls were used for sampling. The annual diameter increment during the period 1971 to 1976 was measured on stem cross sections extracted at breast height, at 50% and at 75% of the total height. The corresponding lengths of the annual apical shoots were also measured. Foliar analysis was carried out for the elements nitrogen, phosphorus, potassium, calcium, and magnesium. In addition, the ovendry weight per 1,000 needles and grams of needles per sample were determined. The latter did not include all needles grown during a given year, but those on the main axis of the branch and of the first side branch. Three humus samples were taken within each experiment plot. The subplots were also used for soil sampling, with three samples taken at random within each subplot.

Growth Responses, 1970-76

Table 3 gives the correlation coefficients between diameter growth during the period 1971 to 1976 at 1.3 m (X_1), at 50% of tree height (X_2), at 75% of the tree height (X_3), and height growth (X_4).

All correlations are highly significant. Diameter growth at 1.3 m is most highly correlated with that at 50% of tree height and drops considerably for the 75% of tree height position. The lowest correlation coefficient is observed for the variables diameter growth, growth at breast height, and height growth. Predicting X_4 from X_1 would leave 85.2% of the total variability of height growth unexplained by X_1.

An analysis of variance with basal area increment at 1.3 m, at 50% and at 75% of the tree height as response variables, produced F-values of 2.5, 3.5, and 2.7 respectively. This showed that the 50% sampling position represents the best position to detect differences between fertilizer means. The mean squares "between fertilizer categories" were 143, 154, and 56 respectively, which seems to indicate that basal area growth of 75% of the tree height responds less strongly to fertilizer

Table 3. Correlation coefficients.

	X_1	X_2	X_3
		value of r	
X_2	0.745*		
X_3	0.537*	0.760*	
X_4	0.385*	0.605*	0.611*

*Significant at $\alpha = 0.01$.

than that at 1.3 m and at 50% of the tree height, although, because of a lower mean square for error, differences between fertilizers remained significant.

An analysis of variance for basal area growth for the three sampling positions combined revealed F-values for position and fertilizer of 36.3 and 7.7 respectively. They indicated highly significant differences among position and fertilizer means, but the interaction position x fertilizer was nonsignificant. The mean basal area increment as a percentage of control representing the means of the three positions of measurement was as follows:

NPKCa/NK	131
LPKCa	106
PKCa/K	112
NKCa/NK	111
NPCa/N	126
NPK/NK	139

The analysis of variance for height growth revealed a highly significant F-value of 3.66, the means expressed as a percentage of control was as follows:

NPKCa/NK	117
LPKCa	90
PKCa/K	114
NKCa/NK	105
NPCa/N	110
NPK/NK	128

The correlation coefficient between relative basal area and height growth was 0.851, indicating significance at $\alpha = 0.05$.

In the analysis of variance for volume increment, the mean square for experimental error was not significantly greater than that for trees within plots (sampling error). Pooling the relevant sources of variability produced a highly significant F-value for testing differences between fertilizer means of 9.8. The results of the a posteriori Newman-Keuls test are given in Table 4.

Differences between the means for control and LPKCa, PKCa, and NKCa as well as those between LPKCa, PKCa, NKCa, NPCa, NPKCa, and NPK are nonsignificant, although such differences are of considerable magnitude. Nonsignificance is primarily attributable to too few sample trees and excessive uncontrolled variability which has reduced the power of the statistical test.

Table 4. Newman-Keuls test for differences among means.

Treatment	Control	LPKCa	PKCa/K	NKCa/NK	NPCa/N	NPKCa/NK	NPK/NK
Mean	24.53	31.79	32.63	32.87	35.07	39.13	41.06
In % of control	100	130	133	134	143	160	167

Foliar Analysis

The seven foliar samples from each of the 105 sample trees represent discrete categories, namely the one-year-old needles from the top whorl and the six age classes on the seventh whorl. The samples were drawn at random from the population of branches of the top and seventh whorls and represent independent samples.

In a univariate analysis of variance of the experimental data, we postulate:

$$X_{ij} = \mu_j + \alpha_j + \varepsilon_{ij}$$

with α_j expressing the group effect and ε_{ij} representing a random variable with mean zero and variance σ^2. The test hypothesis is:

$$H_o : \mu_1 = \ldots\ldots\mu_j = \ldots\ldots\mu_k$$

In the univariate analysis of variance we have:

$$P\ (\text{rejecting } H_o \mid H_o \text{ is true}) = \alpha$$

where

$$\alpha = \text{level of significance}$$

hence

$$P\ (\text{not rejecting } H_o \mid H_o \text{ is true}) = 1-\alpha$$

In the case of p response variables being measured and univariate analyses of variance applied to each of them, the true level of significance is:

$$\alpha^* = \binom{p}{1} \cdot \alpha(1-\alpha)^{p-1} + \binom{p}{2}\alpha^2(1-\alpha)^{p-2} + \ldots\ldots\ \alpha^p$$

$$P\ (\text{not rejecting } H_o \mid H_o \text{ is true}) = 1-\alpha^*$$

For p = 7, we obtain a true level of significance of 0.3017 and the probability of an incorrect conclusion when H_o is true is equal to 0.6983. For this reason, it is more appropriate to consider the vector $\underline{X}$, containing the observations of the p response variables. The assumption will be that the p-dimensional response variable follows the multivariate normal distribution, which is an extension of the univariate distribution and takes care of the higher dimensionality. In other words, the group population means in H_o are now replaced by the vectors of means $\underline{\mu}_1$ .... $\underline{\mu}_j$ .... $\underline{\mu}_k$. The variance σ^2 in the univariate case is replaced by the variance-covariance matrix Σ, in the present case a 7 x 7 matrix, in which the off-diagonal elements, expressing the covariance between the response variables will be nonzero.

The test hypothesis is:

$$H_o: \begin{bmatrix} \mu_{11} \\ \\ \mu_{1p} \end{bmatrix} = \begin{bmatrix} \mu_{21} \\ \\ \mu_{2p} \end{bmatrix} = \dots \begin{bmatrix} \mu_{k_1} \\ \\ \mu_{kp} \end{bmatrix}$$

There are two important criteria which can be used in this multivariate test:

(1) Roy's largest root criterion
Let H be the matrix of the sum of squares and products among groups and E the corresponding error matrix, i.e., the within groups matrix of sum of squares and products. It is then required to calculate the eigenvalues of the matrix product HE^{-1} where E^{-1} = inverse of E. This is equivalent to solving the determinantal equation $| H - \lambda E^{1} |$ = 0.
The largest root is used to calculate $\Theta = \lambda_{largest}/(1 + \lambda_{largest})$.
The critical value for Θ can be obtained from existing tables given by Heck (1960).

(2) Wilk's likelihood criterion

$$\Lambda = \frac{| E |}{| H | + | E |}$$

The statistic

$$R = \frac{1 - \Lambda^{1/s}}{\Lambda^{1/s}} \cdot \frac{ms - p(k - 1)/2 + 1}{p\ (k - 1)}$$

with

$$m = kn - 1 - (p + k)/2$$

$$s = \sqrt{(p^2\ (k - 1) - 4)/(p^2 + (k - 1)^2 - 5)}$$

follows, approximately the F-distribution with $v_1 = p(k-1)$ and $v_2 = ms-p\ (k-1)/2+1$ degrees of freedom.

The results are given in Table 5.

The multivariate tests of significance concerning the equality of the vectors of means is rejected for the elements phosphorus, potassium, calcium, and magnesium, but not for nitrogen.

Table 6, revealing where differences occur, gives the univariate F-values. Phosphorus shows the lowest F-value for needles from the top whorl. For the subsample in the seventh whorl the value of F increases with increasing needle age. Hence, differences among fertilizer means, in relation to random variability, tend to increase with increasing needle age. A similar trend is apparent for the element potassium. The F-values for calcium are invariably high. For magnesium no age trend can be observed. The above F-values cannot be used for tests of significance, but they reveal the sources of observed variability.

Table 5. Largest latent roots and likelihood ratio for each of the elements and probability that observed values are exceeded if H_o is true.

Element	Largest Root	Wilk's Lambda	$P(\Lambda>\Lambda_{obs.})$
Nitrogen	0.863	0.0156	0.05
Phosphorus	0.953	0.0063	0.02
Potassium	0.942	0.0071	0.03
Calcium	0.940	0.0032	0.004
Magnesium	0.945	0.0011	0.0002

Table 7 gives the treatment means for each of the elements with the exception of nitrogen. The nitrogen level for all treatments and needle ages falls below the deficiency limit of 14 mg/g.

In order to assess the time trend apparent in some of the elements, a linear regression equation based on the model $Y_i = \beta_o + \beta_1 X_i + \varepsilon_i$ assuming $\varepsilon(N, 0, \sigma^2)$ was fitted to the six needle age classes of the seventh whorl for each tree separately. An analysis of variance was calculated for each element.

The differences between the mean regression coefficients were significant for the elements potassium (F = 4.3*), calcium (F = 9.1**), and magnesium (F = 6.0**).

Table 6. Univariate F-values.

Response Variable	Nitrogen	Phosphorus	Potassium	Calcium	Magnesium
X_{11}	<1	2.6	1.5	13.2	4.0
X_{71}	<1	7.2	1.6	12.7	1.9
X_{72}	<1	13.6	3.5	21.2	5.9
X_{73}	<1	18.1	6.9	22.7	5.7
X_{74}	<1	31.7	12.8	18.1	9.5
X_{75}	1.0	29.9	16.7	15.3	8.1
X_{76}	<1	34.2	13.2	12.7	8.4

X_{11} : top whorl, one-year-old needles.

X_{71} to X_{76} : seventh whorl, one- to six-year-old needles.

Table 7. Table of means.

Treatment	X_{11}	X_{71}	X_{72}	X_{73}	X_{74}	X_{75}	X_{76}
PHOSPHORUS							
Control	1.32	1.22	1.03	0.98	0.91	0.84	0.78
NPKC	1.46	1.37	1.23	1.13	1.07	0.94	0.83
LPKCa	1.57	1.44	1.27	1.18	1.14	1.06	0.93
PKCa	1.54	1.50	1.36	1.30	1.27	1.18	1.04
NKCa	1.22	1.16	1.04	0.92	0.86	0.75	0.67
NPCa	1.70	1.44	1.34	1.26	1.24	1.09	0.98
NPK	1.38	1.33	1.14	1.06	1.01	0.92	0.82
POTASSIUM							
Control	6.23	5.50	5.47	5.57	5.82	5.94	5.84
NPKCa	5.31	5.93	7.11	5.61	5.48	5.32	4.86
LPKCa	5.79	6.17	5.12	4.69	4.58	4.46	3.96
PKCa	4.57	4.67	5.59	5.45	5.41	5.40	5.16
NKCa	5.40	5.55	5.64	5.42	5.38	5.34	5.08
NPCa	5.66	5.20	4.26	3.98	3.94	3.81	3.71
NPK	5.36	5.51	6.11	6.13	6.21	6.07	5.60
CALCIUM							
Control	4.47	4.71	5.27	6.55	6.04	5.90	6.19
NPKCa	7.46	8.24	9.86	12.82	13.02	12.72	12.18
LPKCa	8.74	9.54	11.98	18.31	19.68	20.10	19.97
PKCa	7.92	8.55	10.46	13.53	14.09	15.26	15.62
NKCa	7.24	8.02	9.79	12.24	11.77	12.26	12.15
NPCa	7.59	8.06	11.58	13.96	14.90	15.41	15.52
NPK	4.86	5.42	5.80	6.80	6.92	7.40	7.14
MAGNESIUM							
Control	1.06	1.09	0.82	0.86	0.80	0.80	0.86
NPKCa	1.30	1.31	1.19	1.19	1.18	1.14	1.09
LPKCa	1.24	1.41	1.39	1.46	1.51	1.51	1.53
PKCa	1.30	1.32	1.19	1.10	1.12	1.17	1.24
NKCa	1.23	1.24	1.10	1.02	0.97	0.97	0.94
NPCa	1.56	1.42	1.17	1.31	1.38	1.11	1.40
NPK	1.42	1.25	0.94	0.82	0.80	0.81	0.78

Similarly, an analysis of variance was calculated for a response variable defined as the product of mean nutrient level in the needles extracted from the seventh whorl multiplied by the ovendry weight of 1,000 needles. Significant differences were observed for the elements phosphorus ($F = 22.8^{**}$), potassium ($F = 4.6^{*}$), calcium ($F = 20.5^{**}$), and magnesium ($F = 7.3^{**}$); but for nitrogen the differences were nonsignificant.

DISCUSSION AND CONCLUSIONS

Foliar analysis did not disclose a statistically significant effect of fertilizer on nitrogen concentration. An analysis of variance also for the response variable nitrogen concentration multiplied by the ovendry weight of 1,000 needles and by ovendry needle weight per sample respectively did not disclose significant differences. This is partly owing to variability between plots within treatments being significantly greater than variability between trees within plots, which may have clouded the effect of fertilizer on nutrient concentration. A more likely explanation is that foliar analysis was carried out seven growing seasons after the application of fertilizer, and differences may have been detectable at an earlier period of foliar analysis. The periodically measured sample plots produced evidence of a strong response of growth rate to the application of nitrogen. It remains uncertain why this was not reflected in the nutrient concentration. Possibly, nitrogen induced an increased production of foliage, but this was not confirmed by the statistical analysis of the study variable "g needles per sample."

There was a significant response of phosphorus concentration to fertilizer. The lightest response occurred in the one-year-old needles of the top and of the seventh whorl, the strongest response in the six-year-old needles of the seventh whorl. A similar trend was observed for the potassium concentration, but the response for calcium was equally strong for both positions and all needle ages. The trend observed for the element magnesium was similar to that for phosphorus, although the univariate F-values were much lower. For the element phosphorus, the univariate F-value tends to increase with increasing needle age, primarily because of a decreasing mean square for error. A similar trend was observed for the element potassium. The mean square "between treatments" for the elements calcium and magnesium, however, tends to increase with increasing needle age. It should be realized that magnesium was not applied as a separate fertilizer application, but in combination with calcium in 1956 and to a lesser degree with the NK application in 1970.

Growth analysis of the 105 sample trees showed convincingly that the study of single tree growth responses instead of those on a unit area basis is inadequate to disclose fertilizer effects. The analysis of variance indicates a rejection of the overall hypothesis of zero differences, but subsequent multiple-range testing produced limited evidence of significance between individual treatment means. This effect is partly the result of growth sampling variability, since ring width at a certain height above ground was determined at not more than two opposite positions along the perimeter of the stem; and it is also partly related to genetic and microenvironmental factors. This has most definitely concealed the occurrence of significant foliar nutrient content-volume growth correlations, which, with the exception of calcium and postassium, were alarmingly low.

REFERENCES

Heck, D. L. 1960. Charts of some upper percentage points of the distribution of the largest characteristic root. Annals of Mathematical Statistics 31:625-642.

Reemtsma, J. B., and A. Van Laar. 1981. The fertilizer trial Oerrel in *Picea abies*, with reference to sampling methodology. XVII IUFRO *World Congress*, Japan, Working Groups S1.02-01 and S1.02-08.

_________. 1982. Auswirkungen der Dügung auf den Kreisflächenzuwachs in verschiedener Stammhöhe Allg. Forst. und Jagdzeitung: 110-115.

FOLIAR RESPONSES CAUSED BY DIFFERENT NITROGEN RATES AT THE REFERTILIZATION OF FERTILE PINE SWAMPS

E. Paavilainen and P. Pietiläinen

INTRODUCTION

Peatlands rich in nitrogen but low in phosphorus and potassium usually benefit from PK fertilization, which leads to a strong and long-lasting growth improvement. On such naturally fertile peat soils, nitrogen application with phosphorus and potassium is recommended only in fully stocked and well-growing forests of old drainage areas (Huikari and Paavilainen 1972; Paavilainen 1979a). The role of nitrogen at refertilization is not adequately known.

When considering the nitrogen requirements at the refertilization of peatlands, attention should be focused on the nutrient ratios. The mere nitrogen refertilization may, on both poor and fertile peatlands, change the nutrient ratios unfavorably and result in a growth decline of stands (e.g., Paavilainen 1977, 1978, 1979b; Kaunisto and Paavilainen 1977; Raitio 1981). Imbalance of both main nutrient and micronutrient concentrations may cause growth disturbances (Huikari 1974, 1977; Kaunisto and Paavilainen 1977; Veijalainen 1975; Raitio 1981). Especially important is to know how the nitrogen rates affect the nutrient ratios and the development of assimilating foliage.

Only one experimental series, carried out on a cotton grass pine swamp, reports on the effect of different nitrogen rates applied with phosphorus and potassium on peatland (Paavilainen 1974). No results dealing with the responses caused by nitrogen rates at the refertilization of fertile peat soils have been published.

The aim of this investigation is to find out how the foliage of stands on fertile pine swamps is affected by nitrogen refertilization in addition to phosphorus, potassium, and micronutrients. The investigated nitrogen rates range from 50 to 600 kg per hectare.

The foliar dry weight, nutrient and chlorophyll levels, and needle structure during two growing seasons after refertilization were studied. Later the observed foliar responses will be compared with the changes in the growth of stands.

MATERIAL AND METHODS

Experimental Fields

The material was collected in the experimental fields of Köhisevä (= K, location 64°5'N, 26°36'E) and Vesikkosuo (= V, location 64°4'N, 26°41'E) established in Bothnia, western Finland, by the Department of Peatland Forestry at the Forest Research Institute. In both areas an experiment was established in a pine seedling stand, a young pine thinning stand, and an old pine thinning stand.

The peatland site type was in Köhisevä *Carex globularis* and sedge pine swamp. In Vesikkosuo experimental field the site types were partly the same as in Köhisevä and partly cotton grass pine swamp.

The seedling stand and the young thinning stand at Köhisevä were first fertilized in 1964 with finely ground rock phosphate (13.8% P) 600 kg/ha and muriate of potash (42% K) 178 kg/ha. The old thinning stand at Köhisevä was fertilized in 1961 with finely ground rock phosphate 600 kg/ha and muriate of potash 200 kg/ha, in addition to which the area was fertilized with 700 kg/ha of NPK (14% N, 7.3% P, 8.3% K) in 1963. The experiments at Vesikkosuo were fertilized with 600 kg/ha of finely ground rock phosphate and 200 kg/ha of muriate of potash in 1962-65.

The eight refertilization treatments were replicated two or three times in all six experiments (see Table 1).

PK fertilizer used was the granular PK for peatland forests which in addition to 8.7% P and 16.6% K among other things contains 0.1% Mg, 17.7% Ca, 2.4% S, 0.5% Na, 1.1% Fe, and 0.2% B. Nitrogen was given as calcium ammonium saltpeter. The micronutrient mixture contained 1.1% B, 12.8% Cu, 5.5% Mn, 9.8% Fe, 5.5% Zn, 1.4% Mo, and 0.7% Na.

Nutrient Analyses

Soil and needle samples were collected in each sample plot before establishing the experiments. The soil sample consisted of five subsamples (4 by 5 cm), and the needle sample of the last needles of the dominant canopy layer (5 trees/plot).

The averages of different nutrients per experiment before refertilization are presented in Table 2. The variation between replications was slight. Only in two cases was it statistically significant according to the analysis of variance (soil nitrogen content in the seedling stand at Köhisevä and foliar nitrogen content in the young thinning stand at Köhisevä).

According to the needle analysis, the nutrient conditions of the stands were good when the experiments were established. Only the foliar phosphorus and potassium levels in young and old thinning stands at Vesikkosuo were rather low (see, e.g., Paarlahti et al. 1971).

The nitrogen content of peat was high in all the experiments. Similarly, the phosphorus content of soil was fairly high, while the potassium and calcium contents were relatively low (see Westman 1981).

Needle samples were taken again from the last needles of the dominant canopy layer in October 1979 (two growing seasons after refertilization). The needles were analyzed for N, P, K, Ca, Mg, B, Cu, and Mo. All nutrient analyses were made by Viljavuuspalvelu Oy (Soil Testing Service Co.).

Chlorophyll Determination

The frozen needle samples were thawed and cut into 1-2 mm sections. 1.0 g of the cut needle mass was put into 25 ml of 80% acetone and homogenized with a Yastral 5000 suspensor for 30 seconds. The green homogenate was filtered through Watman No. 1 filter paper in a Buchner funnel. The cellular debris was washed and refiltered with 25 ml of 80% acetone. The optical density of the filtrate was measured in a 1 cm

Table 1. Fertilization treatments.

Treatment	PK Fertilizer (kg/ha)	Nitrogen (kg/ha)	Micronutrient Mixture
1	---	---	---
2	500	---	---
3	500	50	---
4	500	100	100 kg/ha + 10 kg/ha fertiizer borate (14% B)
5	500	100	---
6	500	200	---
7	500	400	---
8	500	600	---

Table 2. Foliar and soil nutrient levels before refertilization (needle sampling in March 1978 and soil sampling in June 1978).

	Seedling Stand	Young Thinning Stand	Old Thinning Stand
		Köhisevä	
Needles	(C)	(D)	(B)
N (%)	1.54	1.44	1.54
P (%)	0.23	0.19	0.19
K (%)	0.50	0.47	0.42
Soil			
N (%)	2.20	2.20	2.18
P (mg/g)	1.13	1.25	0.79
K (mg/g)	0.20	0.20	0.20
Ca (mg/g)	1.61	1.95	2.86
		Vesikkosuo	
Needles	(B)	(C)	(D)
N (%)	1.50	1.42	1.68
P (%)	0.22	0.16	0.15
K (%)	0.47	0.38	0.39
Soil			
N (%)	2.61	2.29	2.43
P (mg/g)	0.87	0.80	0.86
K (mg/g)	0.13	0.19	0.21
Ca (mg/g)	2.44	2.37	2.98

The standard errors were as follows:
Needles: N = 0.01-0.03%; P = 0.00-0.01%; K = 0.00-0.01%.
Soil: N = 0.03-0.09%; P = 0.02-0.03%; K = 0.00-0.01%; Ca = 0.07-0.15%.

cuvette against a 80% acetone blank at wavelengths 645 and 663 nm with a Hitachi 100-40 spectrophotometer.

Calculations for chlorophyll determinations:

$$\text{mg chlorophyll a/g tissue} = [12.7\ (OD_{663}) - 2.69\ (OD_{645})] \times \frac{V}{1000 \times W}$$

$$\text{mg chlorophyll b/g tissue} = [22.9\ (OD_{645}) - 4.68\ (OD_{663})] \times \frac{V}{1000 \times W}$$

In the above equation, OD is the optical density of the chlorophyll extract of the specific indicated wavelength, V the final volume of the 80% acetone-chlorophyll extract, and W the fresh weight in grams of the tissue extracted (MacKinney 1941).

Needle Structure

The anatomical needle studies were done with a Leitz Dia Lux 20 EB light microscope. The scale in the ocular was calibrated and used to measure the cellular parameters from the cross sections of the needles. The needle sections were cut from the center of the needle by keeping the needle between two microscope slides. The one-and-a-half-cell-thick cross sections were cut with a razor blade. The cross sections were stained with a general anatomical stain (Niemelä et al. 1974).

The anatomical parameters were measured as in Raitio and Rantala (1977) and Raitio (1981). The cell wall thickness and cell diameter were measured by the vascular bundle and at the center of the supportive tissue.

RESULTS AND DISCUSSION

Foliar Dry Weight

The thousand-needle biomass was increased by refertilization (Table 3). PK fertilization alone increased the dry weight of thousand needles, but the best result was obtained by applying nitrogen with phosphorus and potassium. A positive effect was gained even by 50 kg of nitrogen per hectare. Since the nitrogen rate was 600 kg per hectare, the foliar dry weight was, in most cases, lower than on plots where 200-400 kg N/ha were used.

Table 3 also shows that the thousand-needle dry weight decreased as the stand aged. The highest dry weight was in a seedling stand and the lowest in an old thinning stand.

Foliar Nutrient Levels

In two growing seasons some changes occurred in the foliar nutrient levels irrespective of fertilization treatments. On the unfertilized plots the foliar nitrogen levels increased on four and decreased on two out of six experimental fields (Table 4). The foliar phosphorus level decreased on all experimental fields. In the young and old thinning stands at Vesikkosuo in 1979, the phosphorus level was 0.14%, bordering

Table 3. Foliar dry weight two growing seasons after refertilization.

Fertilization			Dry Weight of 1,000 Needles (g)					
			Köhisevä			Vesikkosuo		
N (kg/ha)	Pk fertilizer (kg/ha)	Micronutrient mixture (kg/ha)	Seedling stand	Young thinning stand	Old thinning stand	Seedling stand	Young thinning stand	Old thinning stand
0	0	0	14.79	14.34	13.11	12.98	11.15	11.25
0	500	0	15.10	17.00	17.19	16.58	12.97	13.97
50	500	0	21.47	20.67	19.32	25.62	19.30	12.04
100	500	0	22.53	23.22	18.98	29.58	18.43	16.25
100	500	100 + 1.4 B	24.44	25.07	19.81	26.48	20.98	14.57
200	500	0	26.15	25.19	22.37	28.42	20.11	15.48
400	500	0	28.45	23.64	23.06	27.88	18.01	13.26
600	500	0	25.78	22.45	16.87	32.35	16.46	14.45
F-value and significance			5.63**	3.08*	3.53*	4.56*	2.85	0.85
Significant difference (5% risk level)			10.42	10.91	9.47	17.52	--	--

* = Significant at 0.05 risk level.
** = Significant at 0.01 risk level.

Table 4. Changes in foliar nitrogen, phosphorus, and potassium levels (% of dry weight) occurring between samplings on unfertilized plots.

	Seedling Stand			Young Thinning Stand			Old Thinning Stand		
	N	P	K	N	P	K	N	P	K
					Köhisevä				
March 1978	1.54	0.23	0.50	1.44	0.19	0.47	0.54	0.19	0.42
October 1979	1.55	0.19	0.49	1.33	0.18	0.44	1.62	0.17	0.44
Increase (+) or decrease (-)	+0.01	-0.04	-0.01	-0.11	-0.01	-0.03	+0.08	-0.02	+0.02
					Vesikkosuo				
March 1978	1.50	0.22	0.47	1.42	0.16	0.38	1.68	0.15	0.38
October 1979	1.71	0.19	0.44	1.65	0.14	0.36	1.63	0.14	0.40
Increase (+) or decrease (-)	+0.21	-0.03	-0.03	+0.23	-0.02	-0.02	-0.05	-0.01	+0.01

on phosphorus shortage (see Paarlahti et al. 1979). Few changes occurred in the foliar potassium levels. The potassium level in the young thinning stand at Vesikkosuo had gone down to 0.36%, which calls for potassium fertilization.

The comparison to the nutrient levels after refertilization reveals that PK refertilization increased the foliar phosphorus and potassium levels in all the experiments (Table 5). The foliar nitrogen level increased with higher nitrogen rates. The foliar magnesium level, compared with other main nutrients, decreased in most instances after the application of high nitrogen rates. A statistically significant negative correlation existed between the foliar nitrogen and the other investigated main nutrient levels. Nitrogen fertilization clearly raised the foliar N/P and N/K ratios (Figures 1 and 2).

The optimum N/P ratio of pine needles on peatland has usually been about 11, according to Finnish investigations (see, e.g., Puustjärvi 1962a, 1965; Kaunisto and Paavilainen 1977). Compared with this limit value, nitrogen refertilization with 400 kg N/ha or more (with PK fertilizer) resulted in too high N/P ratios. Only in the sample plots of the seedling stand did the foliar N/P ratio stay around limit value 11 or below it even with high nitrogen rates.

The optimum foliar N/K ratio is 3.5 according to Puustjärvi (1962b, 1965). Compared with this, the nitrogen application of even 200 kg N/ha raised the foliar nitrogen content too high in many sample plots in relation to potassium amounts.

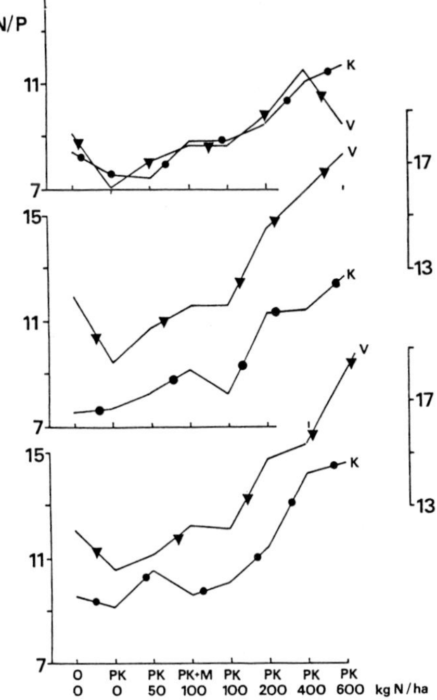

Figure 1. N/P ratio of the needles two growing seasons after refertilization (K = Köhisevä, V = Vesikkosuo).

Table 5. Foliar nutrient levels two growing seasons after refertilization.

Fertilization			Nutrient Content of Needles (% of dry weight)														
			Seedling Stand					Young thinning stand					Old thinning stand				
N (kg/ha)	PK fertilizer (kg/ha)	Micro-nutrient mixture (kg/ha)	N	P	K	Ca	Mg	N	P	K	Ca	Mg	N	P	K	Ca	M
Köhisevä																	
0	0	0	1.55	0.185	0.489	0.230	0.183	1.33	0.176	0.440	0.226	0.167	1.62	0.170	0.440	0.215	0.1
0	500	0	1.61	0.212	0.556	0.233	0.175	1.64	0.208	0.574	0.219	0.173	1.74	0.190	0.556	0.224	0.1
50	500	0	1.59	0.214	0.579	0.220	0.185	1.67	0.201	0.536	0.209	0.164	1.72	0.163	0.511	0.234	0.1
100	500	0	1.83	0.209	0.607	0.258	0.167	1.65	0.200	0.575	0.186	0.165	1.70	0.168	0.460	0.195	0.1
100	500	100+1.4B	1.86	0.212	0.600	0.234	0.170	1.79	0.195	0.550	0.196	0.153	1.74	0.181	0.479	0.205	0.1
200	500	0	2.02	0.212	0.599	0.260	0.148	2.08	0.184	0.487	0.202	0.151	1.92	0.168	0.461	0.210	0.1
400	500	0	2.15	0.194	0.508	0.299	0.147	2.13	0.186	0.466	0.175	0.156	2.41	0.169	0.373	0.198	0.1
600	500	0	2.30	0.196	0.515	0.231	0.140	2.35	0.185	0.456	0.192	0.157	2.35	0.161	0.423	0.178	0.1
F-value and significance			9.67***	1.90	2.47	1.16	3.35*	20.51***	0.96	5.53**	2.15	1.04	12.39**	10.49**	3.43	1.51	6.8
Significant difference (5% risk level)			0.44	--	--	--	0.046	0.37	--	0.116	--	--	0.51	0.017	--	--	0.0
Vesikkosuo																	
0	0	0	1.71	0.189	0.439	0.205	0.175	1.65	0.138	0.357	0.229	0.174	1.63	0.135	0.401	0.220	0.1
0	500	0	1.48	0.208	0.527	0.184	0.161	1.50	0.159	0.507	0.230	0.151	1.79	0.169	0.508	0.227	0.1
50	500	0	1.67	0.208	0.566	0.180	0.189	1.80	0.167	0.500	0.197	0.129	1.72	0.155	0.556	0.208	0.1
100	500	0	1.84	0.211	0.600	0.182	0.171	1.94	0.167	0.556	0.223	0.144	1.86	0.154	0.487	0.217	0.1
100	500	100+1.4B	1.81	0.207	0.577	0.198	0.167	1.95	0.168	0.529	0.202	0.147	1.92	0.156	0.493	0.190	0.1
200	500	0	1.91	0.194	0.550	0.259	0.171	2.36	0.163	0.497	0.182	0.137	2.08	0.141	0.459	0.197	0.1
400	500	0	2.15	0.186	0.530	0.176	0.154	2.40	0.150	0.454	0.167	0.118	2.18	0.142	0.455	0.189	0.1
600	500	0	1.75	0.184	0.530	0.178	0.145	2.67	0.154	0.437	0.172	0.127	2.49	0.138	0.420	0.196	0.1
F-value and significance			2.36	1.34	2.44	2.73	2.85	5.31*	4.16*	4.08*	1.95	4.14*	7.35**	1.86	6.96**	0.70	0.5
Significant difference (5% risk level)			--	--	--	--	--	1.02	0.029	0.179	--	0.050	0.60	--	0.109	--	-

* = Significant at 0.05 risk level.
** = Significant at 0.01 risk level.
*** = Significant at 0.001 risk level.

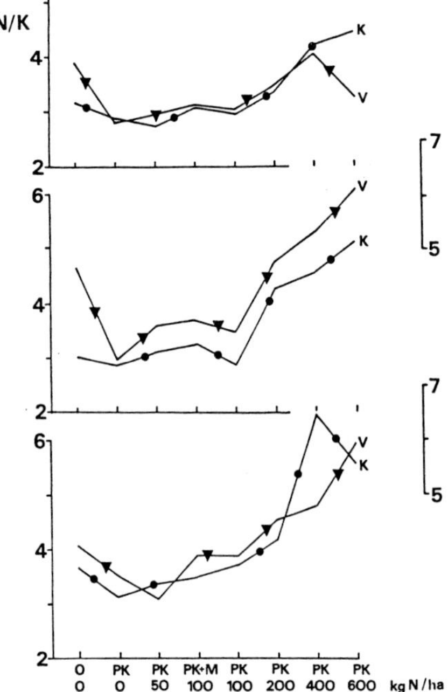

Figure 2. N/K ratio of the needles two growing seasons after refertilization (K = Köhisevä, V = Vesikkosuo).

The boron-containing PK fertilizer alone and especially with micronutrients clearly increased the foliar boron content (Table 6). The highest nitrogen doses slightly decreased the average foliar boron level as compared with the sample plots fertilized with mere PK fertilizer or PK with low nitrogen rates.

The micronutrient fertilization did not affect the foliar copper level much (Table 6). The observed difference in the foliar copper level between the experimental fields (the values were higher in the young and old thinning stands at Vesikkosuo than in the other experimental fields) did not level down during these two growing seasons, although a copper-containing micronutrient mixture was used. In all the experimental fields at Köhisevä and in the seedling stand at Vesikkosuo the copper levels were below 4 ppm, the lowest being 2.6 ppm, which indicates that these experimental fields may suffer from a copper shortage (see, e.g., Kolari 1979; Raitio 1981). The foliar molybdenum level was strongly increased by micronutrient fertilization.

The changes in the foliar nutrient levels after refertilization are clearly shown by the nutrient amounts calculated for a thousand needles (Tables 7 and 8). The amounts also reflect the foliar dry weight increased by fertilization. Thus the total nutrient amounts in needles clearly increased through PK fertilization. Low nitrogen rates (50-100 kg N/ha) further increased the total foliar nutrient amounts. High nitrogen doses did not much contribute to the total nutrient amount, instead it decreased them in some cases.

No distinct correlation between the response to fertilization and foliar nutrient levels by the development stage of stands was observed.

Table 6. Foliar micronutrient levels two growing seasons after refertilization.

Fertilization			Micronutrient Content of Needles (ppm of dry weight)								
			Seedling stand			Young thinning stand			Old thinning stand		
N (kg/ha)	PK fertilizer (kg/ha)	Micronutrient mixture (kg/ha)	B	Cu	Mo	B	Cu	Mo	B	Cu	Mo
Köhisevä											
0	0	0	18.7	3.7	0.057	16.7	3.0	0.065	18.0	2.8	0.046
0	500	0	36.0	3.7	0.052	26.0	3.3	0.057	30.0	3.7	0.044
50	500	0	30.7	3.6	0.050	28.7	3.7	0.070	27.0	2.9	0.024
100	500	0	32.3	3.6	0.055	26.7	3.3	0.051	22.5	3.0	0.032
100	500	100+1.4B	46.3	3.9	0.353	36.7	3.5	0.168	41.0	3.2	0.074
200	500	0	33.3	3.0	0.064	20.7	3.5	0.052	23.5	3.0	0.037
400	500	0	26.3	3.1	0.067	21.3	2.8	0.048	23.0	2.7	0.024
600	500	0	29.3	3.1	0.026	16.3	3.1	0.062	20.5	2.6	0.031
F-value and significance			5.01**	2.13	19.48***	14.19***	2.55	5.25**	12.11**	3.28	1.46
Significant difference (5% risk level)			17.7	--	0.120	9.0	--	0.086	12.0	--	--
Vesikkosuo											
0	0	0	14.5	3.6	0.012	18.5	6.1	0.039	15.0	6.7	0.004
0	0	0	28.5	2.7	0.022	39.0	6.8	0.046	33.5	6.8	0.025
50	500	0	30.0	3.6	0.031	37.5	6.9	0.029	29.5	7.4	--
100	500	0	31.0	3.6	0.036	36.5	7.1	0.010	24.5	6.0	0.102
100	500	100+1.4B	40.5	3.3	0.099	55.0	7.8	0.254	44.5	6.7	0.054
200	500	0	33.5	3.8	0.031	29.0	6.5	0.019	28.0	6.9	0.043
400	500	0	26.0	3.9	0.023	32.5	6.1	0.041	26.0	6.3	0.009
600	500	0	25.5	2.8	0.015	25.5	5.4	0.018	22.0	5.6	0.017
F-value and significance			3.82*	0.51	18.15***	15.93***	2.26	6.94	4.27*	2.12	0.81
Significant difference (5% risk level)			22.2	--	0.038	15.8	--	0.018	24.4	--	--

* = Significant at 0.05 risk level.
** = Significant at 0.01 risk level.
*** = Significant at 0.001 risk level.

Table 7. Main nutrient amounts per 1,000 needles two growing seasons after refertilization.

Fertilization			Nutrient Content of Needles (mg/1,000 needles)														
			Seedling stand					Young thinning stand					Old thinning stand				
N (kg/ha)	PK fertilizer (kg/ha)	Nutrient mixture (kg/ha)	N	P	K	Ca	Mg	N	P	K	Ca	Mg	N	P	K	Ca	Mg
Köhisevä																	
0	0	0	230	28	72	34	27	191	25	63	32	24	212	22	58	28	24
0	500	0	243	32	84	35	26	281	35	98	37	30	298	33	96	39	26
50	500	0	340	46	124	46	39	347	42	111	43	34	332	32	99	45	31
100	500	0	412	46	136	58	37	390	47	135	44	39	322	32	87	37	34
100	500	100+1.4B	455	52	145	57	42	449	50	139	49	38	343	36	98	41	35
200	500	0	524	55	157	68	39	517	48	124	52	38	428	38	102	47	32
400	500	0	614	55	145	65	42	500	45	111	43	38	553	39	86	46	33
600	500	0	595	51	134	60	36	528	42	102	43	35	401	27	72	30	26
F-value and significance			10.25***	5.29**	5.22**	5.75**	2.39	7.76***	1.63	2.47	1.00	1.21	5.40*	3.25	2.30	2.21	3.56*
Significant difference (5% risk level)			2.27	22	65	27	--	214	--	--	--	--	244	--	--	--	13
Vesikkosuo																	
0	0	0	222	25	57	27	23	184	15	40	25	19	182	15	45	25	16
0	500	0	246	34	87	31	27	195	21	66	30	19	251	23	70	31	17
50	500	0	427	53	145	47	48	338	32	95	38	24	206	19	67	26	15
100	500	0	544	62	179	54	52	358	31	103	41	27	295	25	79	36	18
100	500	100+1.4B	479	55	153	52	44	408	35	111	42	31	278	23	71	28	15
200	500	0	545	55	156	71	48	478	33	99	36	28	320	22	71	31	16
400	500	0	601	51	148	49	43	431	27	82	30	21	289	19	60	25	15
600	500	0	565	60	171	57	47	434	25	72	28	21	360	20	61	29	18
F-value and significance			3.45	3.52*	3.93*	5.79*	2.78	5.70*	5.39*	5.14*	2.42	3.98*	3.91*	0.77	1.46	0.27	0.43
Significant difference (5% risk level)			--	39	120	33	--	254	16	58	--	12	165	--	--	--	--

* = Significant at 0.05 risk level.
** = Significant at 0.01 risk level.
*** = Significant at 0.001 risk level.

Table 8. Micronutrient amounts per 1,000 needles two growing seasons after refertilization.

Fertilization			Micronutrient Content of Needles (mg/1,000 needles)								
			Seedling stand			Young thinning stand			Old thinning stand		
N (kg/ha)	PK fertilizer (kg/ha)	Micronutrient mixture (kg/ha)	B	Cu	Mo	B	Cu	Mo	B	Cu	Mo
Köhisevä											
0	0	0	278	55	0.94	239	42	0.93	244	37	0.62
0	500	0	542	55	0.78	435	56	0.93	514	63	0.75
50	500	0	651	77	1.94	593	77	1.41	523	56	0.45
100	500	0	724	82	1.22	622	76	1.26	428	56	0.61
100	500	100+1.4B	1,134	95	8.81	922	88	4.39	818	62	1.48
200	500	0	876	78	1.65	524	89	1.39	517	67	0.84
400	500	0	773	88	1.76	498	66	1.18	528	61	0.54
600	500	0	768	81	0.65	362	69	1.40	352	43	0.53
F-value and significance			3.88*	1.91	8.90***	6.44**	2.38	3.66*	9.97**	2.65	1.51
Significant difference (5%) risk level)			618	--	4.4	392	--	2.9	296	--	--
Vesikkosuo											
0	0	0	191	46	0.31	204	68	0.43	172	76	0.09
0	500	0	479	45	0.36	505	89	0.58	473	96	0.37
50	500	0	770	89	0.79	734	113	0.47	375	89	--
100	500	0	867	108	0.99	670	130	0.35	402	99	1.43
100	500	100+1.4B	1,076	87	2.62	1,151	163	5.32	642	98	0.80
200	500	0	912	113	0.82	591	131	0.39	462	107	0.70
400	500	0	718	112	0.63	588	110	1.55	341	82	0.12
600	500	0	834	91	0.47	418	91	0.28	334	81	0.25
F-value and significance			16.83***	0.74	58.96***	8.14**	2.56	11.69**	1.65	0.37	0.77
Significant difference (5% risk level)			380	--	0.6	537	--	2.8	--	--	--

* = Significant at 0.05 risk level.
** = Significant at 0.01 risk level.
*** = Significant at 0.001 risk level.

Only the foliar P content slightly decreased with the age of the stand, as did the foliar B content. Owing to their greater foliar dry weights, some levels calculated per one thousand needles (e.g., nitrogen, phosphorus, and potassium) were somewhat higher in seedling stands than in young or old thinning stands.

Chlorophyll Contents

There were no significant differences in the foliar chlorophyll contents between the refertilization treatments (Table 9). Thus the chlorophyll content of the needles unexpectedly did not correlate with the foliar nitrogen content. The only two nutrients that had some correlation to the chlorophyll contents were phosphorus and copper. According to the stepwise regression analysis at Köhisevä, foliar phosphorus correlated negatively with the content of chlorophyll a ($F = 8.04$, $R^2 = 0.11$) and chlorophyll b ($F = 19.57$, $R^2 = 0.24$). At Vesikkosuo there was a positive correlation between foliar copper and chlorophyll b ($F = 7.94$, $R^2 = 0.11$).

The chlorophyll contents rose somewhat with the age of the stand (Table 9). The highest chlorophyll a/b ratio (3.34) was observed in the young thinning stands.

Needle Structure

The needle structure did not reflect the changes in the nutrient contents and ratios due to the refertilization. The cell wall thickness of the supportive tissue was hardly at all dependent on the refertilization treatment (Table 10). Neither was there any clear correlation between refertilization and cell diameter (Table 11). These results differ from those in some earlier studies (e.g., Raitio and Rantala 1977; Raitio 1981), where unbalanced fertilization had clearly affected the needle structure of Scots pine. Possibly the needles of older pine stands studied in this investigation have been more tolerant against heavy fertilization than the needles of pine seedlings studied by the aforementioned authors. Also, site types and climatic conditions have been distinctly different in this study compared with the previous ones.

At Köhisevä the cell wall thickness increased but at Vesikkosuo it decreased with the stand age. The cell diameters behaved vice versa. The cell diameter was always larger at the center of the supportive tissue than by the vascular bundle as was observed by Raitio (1981).

Table 9. Foliar chlorophyll a and b contents two growing seasons after refertilization.

Fertilization			Chlorophyll Contents								
			Seedling stand			Young thinning stand			Old thinning stand		
N (kg/ha)	PK fertilizer (kg/ha)	Micronutrient mixture (kg/ha)	chl a	chl b	total	chl a	chl b	total	chl a	chl b	total
						Köhisevä					
0	0	0	0.270	0.092	0.362	0.370	0.114	0.484	0.539	0.183	0.722
0	500	0	0.250	0.080	0.330	0.393	0.115	0.508	0.484	0.160	0.644
50	500	0	0.235	0.091	0.326	0.364	0.109	0.473	0.472	0.143	0.615
100	500	0	0.289	0.097	0.386	0.410	0.127	0.537	0.482	0.161	0.643
100	500	100+1.4B	0.276	0.087	0.363	0.434	0.130	0.564	0.533	0.180	0.713
200	500	0	0.257	0.085	0.342	0.393	0.120	0.513	0.547	0.175	0.722
400	500	0	0.288	0.098	0.386	0.390	0.115	0.505	0.479	0.156	0.635
600	500	0	0.316	0.115	0.431	0.372	0.108	0.480	0.454	0.142	0.596
						Vesikkosuo					
0	0	0	0.281	0.096	0.377	n.d.	n.d.	n.d.	0.360	0.139	0.499
0	500	0	0.284	0.100	0.386				0.360	0.148	0.507
50	500	0	0.303	0.108	0.411				0.364	0.143	0.507
100	500	0	0.450	0.159	0.609				0.412	0.154	0.566
100	500	100+1.4B	0.292	0.104	0.396				0.381	0.159	0.540
200	500	0	0.310	0.104	0.414				0.357	0.132	0.489
400	500	0	0.405	0.144	0.549				0.351	0.143	0.493
600	500	0	0.415	0.134	0.549				n.d.	--	--

Table 10. Cell wall thickness in the supportive tissue two growing seasons after refertilization.

Fertilization			Cell Wall Thickness of Supportive Tissue (μm) (1) by the vascular bundle (2) at the center of supportive tissue					
			Seedling stand		Young thinning stand		Old thinning stand	
N (kg/ha)	PK fertilizer (kg/ha)	Micronutrient mixture (kg/ha)	1	2	1	2	1	2
			Köhisevä					
0	0	0	7.3	6.7	7.0	6.6	7.9	7.7
0	500	0	6.9	5.7	7.7	6.3	8.7	6.8
50	500	0	6.3	6.0	8.1	7.0	8.4	7.2
100	500	0	6.6	5.8	8.1	6.2	8.5	7.4
100	500	100+1.4B	6.4	5.5	7.8	6.6	7.7	7.2
200	500	0	6.4	5.5	8.1	5.4	7.9	6.9
400	500	0	6.8	5.4	6.5	5.5	8.2	7.4
600	500	0	6.7	5.5	7.5	5.6	7.6	6.4
			Vesikkosuo					
0	0	0	7.6	7.5	6.5	7.0	5.4	5.9
0	500	0	8.8	7.8	6.2	6.4	6.3	6.2
50	500	0	8.4	6.6	5.2	4.4	5.9	6.1
100	500	0	8.4	7.4	5.5	5.1	7.4	7.3
100	500	100+1.4B	8.9	8.4	5.4	5.8	5.8	6.2
200	500	0	8.1	5.7	6.9	7.1	5.8	5.7
400	500	0	8.3	6.9	5.9	5.9	5.8	5.4
600	500	0	8.3	5.6	5.1	5.7	--	--
			4.93	9.73	8.61	2.52	4.19	4.47

Table 11. Cell wall diameter in the supportive tissue two growing seasons after refertilization.

Fertilization			Cell Diameter of Supportive Tissue (μm) (1) by the vascular bundle (2) at the center of supportive tissue					
N (kg/ha)	PK fertilizer (kg/ha)	Micronutrient mixture (kg/ha)	Seedling stand		Young thinning stand		Old thinning stand	
			1	2	1	2	1	2
			Köhisevä					
0	0	0	26.0	28.3	20.3	25.0	20.8	23.0
0	500	0	24.4	30.4	21.5	25.6	21.9	23.1
50	500	0	26.1	32.3	21.3	26.8	21.6	24.8
100	500	0	26.1	29.8	22.0	28.1	22.3	25.0
100	500	100+1.4B	23.9	28.1	21.2	27.3	20.7	24.0
200	500	0	27.7	30.8	21.8	25.4	21.6	24.5
400	500	0	26.3	29.3	25.1	27.6	20.4	24.1
600	500	0	27.3	30.7	21.4	25.8	20.7	22.1
			Vesikkosuo					
0	0	0	20.9	23.8	24.1	25.5	26.6	27.2
0	500	0	23.2	27.2	24.9	28.4	25.0	26.3
50	500	0	22.3	25.7	24.9	27.4	25.9	26.4
100	500	0	21.5	26.4	24.4	26.0	24.3	26.6
100	500	100+1.4B	22.7	26.1	25.2	26.8	25.2	26.3
200	500	0	21.3	26.9	23.5	25.9	24.2	25.0
400	500	0	21.1	25.5	25.6	27.8	24.1	24.1
600	500	0	21.3	26.6	24.2	26.1	--	--

REFERENCES

Huikari, O. 1974. Hivenravinteet ja puiden kasvu. Metsä ja Puu 11:24-25.

_________. 1977. Micro-nutrient deficiencies cause growth disturbances. Silva Fenn. 11(3):251-255.

Huikari, O., and E. Paavilainen. 1972. Metsänlannoitus. 2. painos. Helsinki. 68 pp.

Kaunisto, S., and E. Paavilainen. 1977. Response of Scots pine to nitrogen refertilization on oligotrophic peat. Commun. Inst. For. Fenn. 92(1):1-54.

Kolari, K. 1979. Hivenravinteiden puute metsäpuilla ja männyn kasvuhäiriöilmiö Suomessa: Kirjallisuuskatsaus. (Micronutrient deficiency in forest trees and dieback of Scots pine in Finland: A review.) Folia For. 389:1-37.

MacKinney, D. 1941. Absorption of light by chlorophyll solutions. J. Biol. Chem. 140:315-322.

Niemelä, T., M. Pyykkö, and M. Uotila. 1974. Mikrotekniikan kurssi. Helsingin yliopiston kasvitieteen laitoksen monisteita 14:1-38.

Paarlahti, K., A. Reinikainen, H. Veijalainen. 1971. Nutritional diagnosis of Scots pine stands by needle and peat analysis. Commun. Inst. For. Fenn. 74(5):1-58.

Paavilainen, E. 1974. The use of nitrogen in fertilizing peatland forests. Proc. Int. Symp. Forest Drainage. Jyväskylä-Oulu, Finland, pp. 337-345.

_________. 1977. Jatkolannoitus väharavinteisilla rämeillä: Ennakkotuloksia. (Refertilization on oligotrophic pine swamps: Preliminary results.) Folia For. 327:1-32.

_________. 1978. PK-lannoitus Lapin ojitetuilla rämeillä: Ennakkotuloksia. (PK fertilization on drained pine swamps in Lapland: Preliminary results.) Folia For. 343:1-17.

_________. 1979a. Metsänlannoitusopas. Helsinki. 112 pp.

_________. 1979b. Jatkolannoitus runsastyppisilä rämeillä: Ennakkotuloksia. (Refertilization on nitrogen-rich pine swamps: Preliminary results.) Folia For. 414:1-23.

Puustjärvi, V. 1962a. Turpeen typen mobilisoitumisesta ja sen käyttökelpoisuudesta suometsissä neulasanalyysin valossa. (On the mobilization of nitrogen in peat and its usefulness in peatland forests in the light of needle analyses.) Suo 13(1):2-10.

__________. 1962b. Suometsien fosforiravitsemuksesta ja neulasten P/N-suhteesta neulasanalyysin valossa. (On the phosphorus nutrition of wet peatland forests and on the P/N ratio in their needles.) Suo 13(2):21-24.

__________. 1965. Neulasanalyysi männyn lannoitustarpeen ilmentäjänä. (The analysis of needles as an exponent for the need of fertilization of Scots pine.) Metätal. Aikakausilehti 1:26-28.

Raitio, H. 1977. Tallarnas växtstörningar, markens näringsbalans och mikronäringsbrist. Silva Fenn. 11(3):255-257.

__________. 1981. Pääravinnelannoituksen vaikutus männyn neulasten rakenteeseen ja varinnepitoisuuteen ojitetulla lyhytkorsinevalla. (Effect of macronutrient fertilization on the structure and nutrient content of pine needles on a drained short sedge bog.) Folia For. 456:1-10.

Raitio H., and E. M. Rantala. 1977. Männyn kasvuhäiriön makro- ja mikroskooppisia oireita: Oireiden kuvaus ja tulkinta. (Macroscopic and microscopic symptoms of a growth disturbance in Scots pine: Description and interpretation.) Commun. Inst. For. Fenn. 91(1):1-30.

Veijalainen, H. 1975. Kasvuhäiriöistä ja niiden syistä metsäojitusalueilla. (Dieback and fertilization on drained peatlands.) Suo 26(5):87-92.

Westman, C. J. 1981. Fertility of surface peat in relation to the site type and potential stand growth. Acta For. Fenn. 172:1-77.

20

FERTILIZERS FOR FOREST USE IN JAPAN

M. Hamamoto

DEVELOPMENT OF FOREST FERTILIZATION IN JAPAN

The forest area in Japan is 25 million hectares, covering two-thirds of the whole land. One-third of the area is national forests and the other two-thirds are private forests. Man-made forests are established in about 30% of the area forming national forests and about 40% of private forests.

Figure 1 shows the fertilized forest area per annum estimated from the quantities of fertilizer used in forest land. Fertilization was introduced in forest management in 1960, mostly for newly planted young trees, and it increased year by year during the 1960s. Fertilized areas reached a high of 94,000 hectares in 1970. After that, with the decrease of annual afforestation area and the economic changes in management of national forests after the oil crisis, the fertilized area decreased and stayed at 55,000 to 60,000 hectares annually in 1975-80. This area corresponds to 0.2% of total forest area and 25% of annual planted area; but the habitually fertilized area comes to 11% to 12% of annual planted

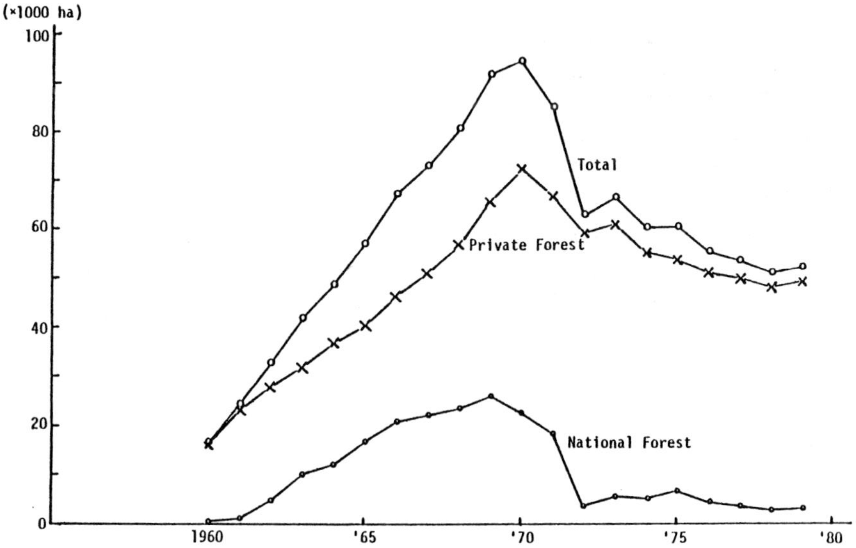

Figure 1. The fertilized area of forests per annum in Japan.

area, taking into account two or three successive years of fertilizing the same stands.

Fertilization practices vary in districts according to the sites and stands. Figure 2 shows the ratio of fertilized area to newly planted area of private forests in 1976, dividing whole forests into eight districts. We can see that fertilization is more habitual in southwestern districts, where the climate is rather warm with more precipitation, supporting the growth of valuable sugi (Cryptomeria japonica) and hinoki (Chamaecyparis obtusa), than in the colder northern districts, where relatively lesser economic species (Abies mayriana, Picea jezoensis, and Japanese larch) are mostly planted. The small ratio of fertilized area in Chugoku districts, where the main soil is developed from weathered granite and Japanese red or black pine are mostly planted, shows the small response to fertilizer economically.

Yoshida (1979) estimates that roughly 70% of fertilized forests are sugi, 25% are hinoki, and 5% others. He also forecasts that the fertilization for Quercus acutissima will increase in future because these logs are in demand for breeding Shiitake mushrooms (Cortinellus shiitake).

KINDS OF FERTILIZERS FOR FOREST USE IN JAPAN

Since the forest soils in Japan are generally low in productivity but not extremely deficient in a particular nutrient, compound fertilizers containing nitrogen, phosphorus, and potassium in good proportion have been used. The higher-analysis compound fertilizers are preferred in mountain forests. Over half the consumed fertilizers, amounting to 24,000 to 30,000 tons per year, are 20-10-10 fertilizers manufactured from urea, ammonium phosphate, and potassium chloride.

Compound fertilizers containing specific ingredients such as dicyandiamide as a nitrification inhibitor, Frenock (sodium tetra fluoro propionate) as a herbicide for Japanese pampas grass or bamboo grass, and peat moss as an organic matter source are on the market even though they are used in a small amounts.

Ammonium nitrate is scarcely used in Japan because of its large hygroscopic property and leaching loss. Calcium cyanamide is considered a good nitrogen fertilizer because of its slow acting and weeding effect.

Since most forest land is covered with volcanic ash soil and its phosphorus absorption capacity is very great, specific phosphatic fertilizers unique to Japan, such as YORIN and LINSTAR, are expected to enhance the absorption ratio of phosphorus by trees. The former is manufactured by a dry process and the latter by a wet process. These phosphatic fertilizers are water insoluble but soluble in citric acid and are effective for longer periods, because of slow fixation by active aluminum and iron in volcanic ash. These fertilizers not only contain phosphorus and calcium but also effective magnesium and silica.

The slow-release nitrogen fertilizers are vaguely expected by many foresters to raise the efficiency of nitrogen. Two types of fertilizers containing slow-release nitrogen are on the market in Japan. One is a compound fertilizer containing Ureaform (a condensation product of urea and formaldehyde) in a ratio of one-fourth to one-half of total nitrogen in the compound fertilizer. The other is a specially formulated large-size IBDU (isobutylidene diurea, a condensation product of urea and isobutyraldehyde), called WOODACE, to maintain effectiveness for three or

District

① Hokkaido

② Tohoku

③ Kanto

④ Hokuriku, Tokai, and Koshin

⑤ Kinki

⑥ Chugoku

⑦ Shikoku

⑧ Kyushu and Okinawa

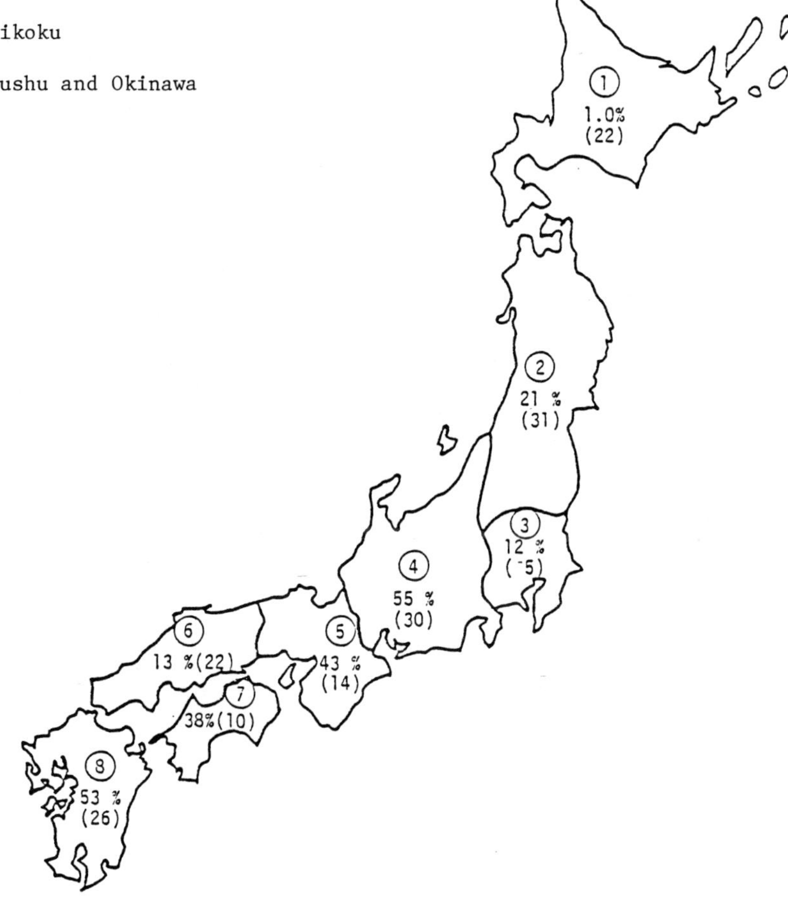

Figure 2. Ratio of fertilized area to newly planted area of private forest in 1976.

%, Ratio ( ), Newly Planted Area (x 1,000 ha)

Table 1. Main chemical fertilizers for forest use in Japan.

Nutrient Content (%) $N - P_2O_5 - K_2O$	Main Constituents
20 - 10 - 10 20 - 12 - 12 24 - 12 - 12	Urea, ammonium sulphate, ammonium phosphate, and potassium chloride
20 - 10 - 10 20 - 12 - 12 24 - 16 - 11 22 - 10 - 10 14 - 18 - 16	Urea, ammonium phosphate, potassium chloride, and Ureaform
10 - 5 - 5	Calcium cyanamide, superphosphate, and potassium chloride
18 - 8 - 9	Urea, ammonium phosphate, potassium chloride, DD as nitrification inhibitor
20 - 8 - 8	Urea, ammonium phosphate, potassium chloride, and Frenock as herbicide
6 - 4 - 3 12 - 8 - 6 6 - 12 - 8 15 - 8 - 8	Ammonium sulphate, superphosphate, YORIN, and potassium chloride
10 - 10 - 10 - 2(MgO)	Urea, IBDU, superphosphate, YORIN, and potassium chloride
23 - 2 - 0 15 - 8 - 0 - 4(MgO) 12 - 6 - 6	IBDU, YORIN or LINSTAR, and potassium silicate are briquetted
17 - 10 - 10	Urea, ammonium phosphate, and potassium chloride are coated with resin

more years after fertilization. This is used mostly for young tree planting.

Since there is much to discuss concerning use of slow-release nitrogen fertilizers for forest land, the subject is treated in a later section of this article.

Table 1 shows the kinds of fertilizers used in the forest, cited from _Hiryo (Fertilizer) Hand Book for Foresters_ by Shibamoto (1979).

PRESENT FERTILIZATION PRACTICES IN JAPAN

Fertilization is used mostly for young trees being planted, and partly for closed-canopy middle-aged forests. The former practices conducted in many forests during the past twenty years show that fertilization has increased the growth of trees by about 50%. For example, results showed height growth of 149 ± 48% from 117 experiments for sugi, 147 ± 64% from 78 experiments for hinoki, and 126 ± 27% from 38 experiments for Japanese pine.

These effects give foresters the advantage of costing down weeding in the newly planted forests, where growth of weeds is normally vigorous. The usual practice is three successive fertilizations (one each year) at an application rate of 10 grams of nitrogen per tree annually, with phosphorus and potassium. Improved products, such as slow-release nitrogen fertilizers, are desired to avoid salt damage, minimize leaching loss, and maintain long-lasting effects.

For closed-canopy middle-aged forests, the effects of fertilization vary according to forest soils, sites, and tree species. From the data of experiments under the best working systems--including pruning, thinning, and fertilization--Kawana (1979), Shibamoto (1979), and Yoshida (1979) have suggested the possible and expected growth curves of trees (see Figure 3).

The Forest Fertilization Society of Japan has proposed a fertilization schedule for sugi to increase the healthy growth of trees and their volumetric yield, as shown in Table 2.

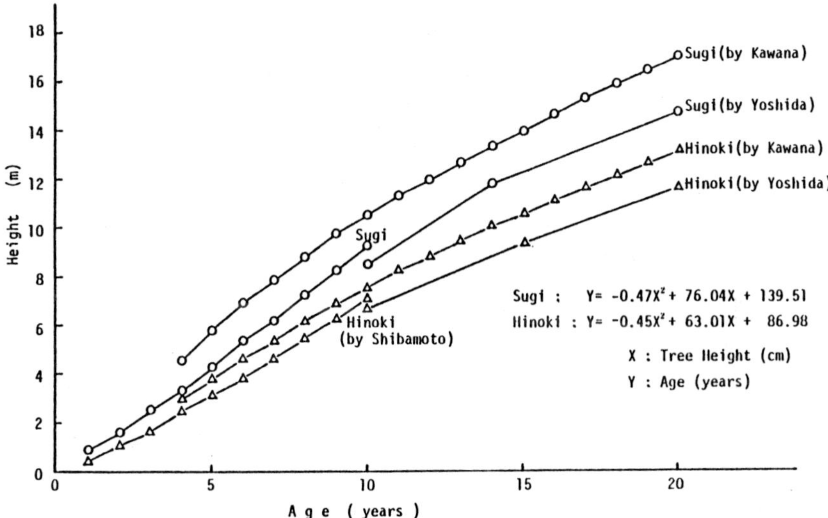

Figure 3. The possible and expected growth curve of trees (Kawana 1979; Yoshida 1979; Shibamoto 1979).

SOME PROBLEMS OF USING SLOW-RELEASE NITROGEN FERTILIZERS

Slow-release nitrogen fertilizers have long been expected to provide nitrogen constantly over a long period for trees in middle-aged forests or for saplings in young plantations without any salt damage. They are also expected to be a better nitrogen fertilizer in the tropical and subtropical rain forests where loss of water-soluble nitrogen applied is

Table 2. Recommended fertilization schedule for sugi (*Cryptomeria japonica*) in middle-aged site (Forest Fertilization Society of Japan).

	Indicator									Fertilization			
	Height		Diameter									
Age (yrs)	Mean (m)	Increase (m/year)	Mean (cm)	Increase (cm/year)	Form Exponent	Number of trees (trees/ha)	Volume (m^3/ha)	Working System	Schedule	N (kg/ha) [g/tree]	P_2O_5 (kg/ha)	K_2O (kg/ha)
0	0.5											
1	1.0	0.5				3,500		planting, weeding	1st	30 [9]	10-20	10-20
2	1.8	0.8				(3,200)		weeding	2nd	40 [13]	15-25	15-25
3	2.6	"				"						
4	3.4	"				"			3rd	60 [19]	20-40	20-40
5	4.2	"	5.0		84	(3,000)						
6	5.0	"	6.2	1.2		2,700		salvage cutting (300-500	4th	60 [22]	20-40	20-40
7	5.7	0.7	7.2	1.0		"		trees/ha)				
8	6.4	"	8.2	"		"		pruning (<1.5 m)				
9	7.1	"	9.2	"		"						
10	7.8	"	10.2	"	76	"						
11	8.4	0.6	11.1	0.9		"		pruning (<3 m)	5th	80 [30]	30-50	30-50
12	9.0	"	12.0	"		"						
13	9.6	"	12.9	"		"						
14	10.2	"	13.8	"		1,800	185	thinning (900 trees, 33% 32 m^3/ha, 15%)				
15	10.8	"	14.7	"	73	"		pruning (<5 m)	6th	100 [56]	15-50	15-50
16	11.4	"	15.6	"		"						
17	12.0	"	16.5	"		"						
18	12.5	0.5	17.4	"		1,250	245	thinning (500 trees, 31%				
19	13.0	"	18.2	0.8		"		32 m^3/ha, 12%)	7th	150 [120]	20-75	20-75
20	13.5	"	19.0	"	71	"						
21	14.0	"	19.8	"		"						
22	14.5	"	20.6	"		"						
23	15.0	"	21.3	0.7		"						
24	15.5	"	22.0	"		900	325	thinning (350 trees, 28%				
25	16.0	"	22.7	"	70	"		32 m^3/ha, 9%)	8th	200 [222]	30-100	30-100

Table 2. Cont'd.

	Indicator								Fertilization			
	Height		Diameter									
Age (yrs)	Mean (m)	Increase (m/year)	Mean (cm)	Increase (cm/year)	Form Exponent	Number of trees (trees/ha)	Volume (m^3/ha)	Working System	Schedule	N (kg/ha) [g/tree]	P_2O_5 (kg/ha)	K_2O (kg/ha)
26	16.5	"	23.4	"		"						
27	16.9	0.4	24.0	0.6		"						
28	17.3	"	24.6	"		"						
29	17.7	"	25.2	"		"						
30	18.1	"	25.7	0.5	70	"	430					
31	18.5	"	26.2	"		"						
32	18.9	"	26.7	"		"						
33	19.3	"	27.2	"	70	"	473	final cutting				
Total	Mean	0.58	Mean	0.82				Total thinning 96 m^3/ha	8 times	720 [491]	160-140	160-140

great. What follows is a discussion of the problems and possibilities of using slow-release nitrogen in forest land considering the fundamental features of each fertilizer.

Relation Between Quantity Applied and Long-lasting Effect

A fundamental feature of slow release nitrogen fertilizers is that the slower the release rate the more fertilizer is needed in a single application to provide nitrogen for the normal growth of trees. See De Ment (1961), and Figure 4. De Ment's experiment was conducted with Oxamide and clearly showed that the larger the particle size of Oxamide (meaning the slower the release rate) the more fertilizer is needed.

If we broadcast slow-release nitrogen fertilizer on the surface of the forest land using the same nitrogen quantities as in conventional water-soluble nitrogen fertilizer, the trees will be hungry for nitrogen. Too little available nitrogen will result in no visible effects of the fertilization. But the application of large amounts of these rather expensive fertilizers is not feasible, even though they act effectively for long periods.

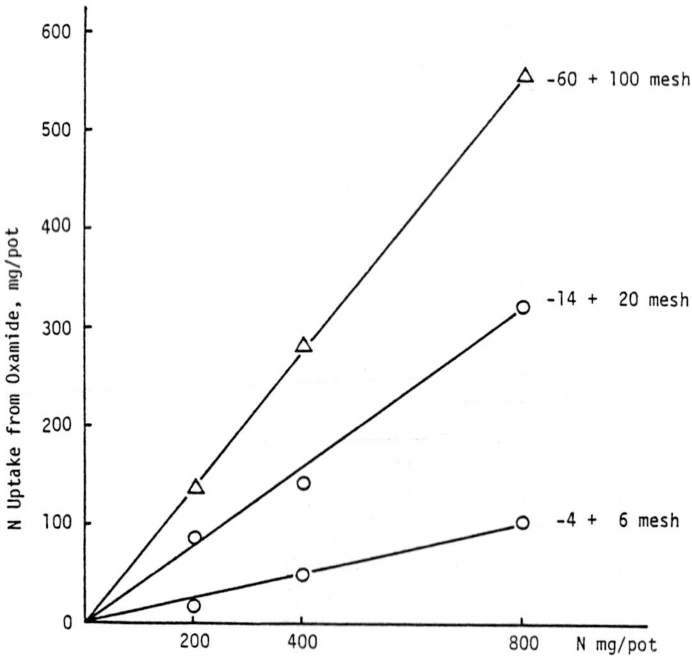

Figure 4. Effect of granule size and rate of application on uptake of nitrogen from Oxamide (De Ment 1961).

Hartsells fine sandy loam
Average for unlimed, pH 5.2, limed, pH 7.3

At present, the broadcasting of slow-release fertilizers cannot be adopted in forest management practices. We must wait for the production of low-priced slow-release fertilizers if we want to use them this way.

Use of Slow-release Nitrogen Fertilizer at Planting Time

Since slow-release nitrogen fertilizers are generally safe from the danger of burning with heavy application, they are better fertilizers for the weak roots of saplings. The rhizosphere of a newly planted young tree is small compared with that of a middle-aged tree, so it is possible to maintain a high level of available nitrogen in a young tree's rhizosphere by putting an adequate amount of slow-release fertilizer in the planting hole. This is an effective use of slow-release fertilizers. Furthermore, it is desirable to apply better fertilizers whose effects last longer with a single application and whose use makes young trees grow faster, thus saving weeding costs.

The Kinds of Slow-release Nitrogen Fertilizer

The term "slow-release nitrogen fertilizer" covers many kinds of fertilizers having different raw materials, processing methods, periods of effectiveness, and mineralization mechanisms. The words "slow-release" and "slow acting" are also used confusingly. The following are brief descriptions of fertilizers now on the world market.

Nitrification inhibitors added to the conventional water-soluble nitrogen fertilizers such as ammonium salts or urea have the effect of inhibiting biological oxidation of ammonium to nitrate. N-Serve in the United States and AM and DD in Japan are known as representatives of this kind (chemical configuration: pyridine, pyrimidine, and dicyandiamide, respectively). Triazine series are also known to be effective.

These are very interesting products if they work as anticipated in laboratories, and to do so they must be mixed thoroughly with soils or they must move together with ammonium ions in the soils. But in practical use they are very difficult to anticipate. What we can do is consider the physical characteristics such as vapor pressure or water solubility so that we can obtain the desired effects. These fertilizers containing inhibitors cannot avoid salt concentration damage, because they are not improved to the point of lowering the water solubility of each fertilizer.

Conventional water-soluble nitrogen fertilizers coated with membrane that is water insoluble but slightly water permeable are also promising products, and SCU (sulfur-coated urea) of T.V.A. in the United States is well know. Various membranes other than sulfur are also under development. These are used in agriculture and horticulture, but further improvements are needed to prolong the duration of effectiveness for forest use. The fertilizers of this group should be called "slow release" because they release the available nutrient itself, in contrast to the next group discussed, the "slow acting."

Slightly water soluble nitrogen compounds such as UF (Ureaform), CDU (crotonylidene diurea), IBDU (isobutylidene diurea), and Oxamide (oxalic acid amide) are now on the world market. These fertilizers first dissolve into soil water very slowly because of their slight water solubility. They diffuse into every part of the soil as inactive forms and then decompose to available forms by the action of soil microorganisms or chemical hydrolysis. UF and CDU decompose by the former action, and IBDU, Oxamide, and smaller molecular parts of UF decompose by both actions. These fertilizers should be called "slow acting" because this description lets us consider the mechanism of mineralization.

The former group will act effectively in the microbiologically active soil, but when the soil microbial activity is poor because of peculiar soil conditions or low temperature, these fertilizers may be ineffective and remain almost void in the forest land or leach away from the rhizosphere of trees.

The latter group whose mineralization depends not only on biological decomposition but on chemical hydrolysis is always effective, even though the period of effectiveness fluctuates with the soil temperature, acidity, or water content of the soil.

Furthermore, the period of effectiveness of the slightly water soluble nitrogen fertilizers is easily controlled by adjusting the size and hardness (even in water) of fertilizer granules as seen in the briquetted IBDU, Ureaform, and Oxamide, as reported by Hamamoto (1976), Koberg (1976), and De Ment (1961), respectively.

Future Research on Fertilizer

Forestry (biomass breeding in a wide sense) is now in the limelight in the expectation of solar energy storage and future raw materials for chemical industries, as well as for paper and timber. Forest fertilization, especially the application of nitrogen fertilizer, has been used to increase forest productivity generally and to expand man-made forests. In northern temperate climates, even simple nitrogen fertilizers such as ammonium nitrate and urea are being used, but in tropical and subtropical rain forests, despite plentiful solar energy, high temperatures, and rainfall suitable for plant growth, the application of water-soluble nitrogen proves insufficient because of leaching loss.

Urea aldehyde condensates, which are the typical slow acting nitrogen fertilizers, show the applicability for advanced fertilizer management in forestry, but the applicability has decreased because of the tight supply of raw materials for condensates since the oil crisis. The development of C_1 chemistry as a new systematic technology based on C_1 compounds is expected to solve the problem of raw materials supply, and then Oxamide and Ureaform using raw materials that come from C_1 compounds or their derivatives might be nominated as the future fertilizers for forest use. In order to increase their applicability, it is necessary (1) to develop a better production technique and (2) to improve the Activity Index (A.I. value) in Ureaform.

Furthermore, it is desirable to promote joint research work in development of new fertilizers and development of their application techniques in biomass fields for future successful forest fertilization.

A Way of Promoting Research

Chemical fertilizers have advanced world agriculture most effectively. The chemical fertilizers now on the market are the products of research and development by the chemical industries and chemical engineering companies. New products such as slow-release or slow-acting fertilizers are also examples of their activities. No one except these companies has the ability to develop a new chemical fertilizer. But they have aimed their activities at profitable markets such as horticulture and have not tried to find a fertilizer suitable for forest use, since they have not thought it profitable.

Unfortunately, foresters have not had the opportunity to participate in research and development to create a new fertilizer for forests. They have only studied how to use available fertilizers. For forest areas they have been selecting the cheapest fertilizer, and have had to settle for low efficiency. No one has ever tried to find a specific fertilizer for forestry.

In order to develop a successful fertilizer for forests, we must conduct wide fertilizer testing, with promising trial products supplied through the cooperation between the chemical companies and forestry research organizations. This project must proceed against the realization that people will most probably not use an expensive fertilizer, even if it is an improved one. There must be cooperative research work among all research organizations in the future.

REFERENCES

De Ment, J. D. 1961. Hydrolysis, nitrification and nitrogen availability of Oxamide, as influenced by granule size. Agricultural and Food Chemistry 9(6):453-456.

Hamamoto, M. 1976. Exploratory trial for betterment of fertilization at tree planting. XVI IUFRO World Congress, Norway. Voluntary paper.

Kawana, A. 1979. Soreirin hibai ni tsuite. (Fertilization for established forest.) Nihon Ryuan Kogyo Kyokai (Association of Japan Ammonium Sulphate and Urea Industry).

Koberg, H. 1976. Eine neue forst-dungetablette. XVI IUFRO World Congress, Norway. Voluntary paper.

Shibamoto, T. 1979. Hiryo (fertilizer) Hand Book for Foresters.

Yoshida, I. 1979. Rinchi hibai no genjo ni tsuite. (The present forest fertilization in Japan.) Nihon Rinchi Hibai Kyokai (Forest Fertilization Society of Japan).

21

STUDIES ON PINE STAND FERTILIZATION. I. SOIL TESTING RESULTS AND DISCUSSION

A. Ostrowska

INTRODUCTION

Generally speaking, soil testing includes soil sampling and analyzing for available nutrients. The testing results are then discussed and recommendations for fertilization are suggested. In the opinion of several authors, numerous problems need to be solved before soil testing can be applied to forestry (Burg 1976; Kowalkowski 1973; Ostrowska 1977; Pritchett 1979). In this connection, investigations were carried out to determine the fertility of soils under pine stands.

CHARACTERIZATION OF SOILS AND STANDS UNDER STUDY

Investigations were carried out in thirty pine stands in different parts of Poland, including sixteen fresh-site stands, seven dry-site stands, and seven fresh-mixed-site pine stands. In general, the age of the pines ranged from 40 to 60 years. In each stand, the soil profiles were analyzed to determine the physical and chemical properties. In particular, the total content in the soil of nitrogen, phosphorus, potassium, magnesium, and calcium was defined. In most cases, the results have shown that under pine stands on dry and fresh sites there are rusty podzolized soils. This type of soil has the following genetic horizons: Ofh (raw humus), AE (accumulation-eluvial), BvJ (weathering-illuvial), Bv (weathering), and C (parent rock). These soils consist of sands of several geological formations, and their composition is generally loose or slightly loamy. The basic element of these soils is quartz (SiO_2, about 90-98%). The soils are acid (pH 3 to 4.5), especially in the upper genetic horizons. They have a high content of free aluminum as well as a low exchange capacity to basic cations--often less than 1 m/100 g of soil (Król and Ostrowska 1980; Ostrowska 1981). They also have a very low water-holding capacity (Król and Trześniewska 1978). This kind of soil is typical under pine stands in Poland.

It should be emphasized that in the individual pine stands under study, a homogeneity of soil types was observed as well as a slight variability in soil properties. In our climatic and soil conditions the predominant mass of roots in the mature pine stands is above 30 to 40 cm depth, which, in most cases, corresponds to the Ofh, AE, and BvJ genetic horizons.

MEAN FERTILITY OF SOILS

It has been established that the results of analyzing the composite samples taken from the individual genetic horizons of the soil root layer constitute a good characterization of the average soil fertility under the pine stands (Ostrowska 1977).

In the present study the composite samples were collected from the Ofh, AE, and Bv (BvJ + Bv) horizons separately, in every stand. Each composite sample was from fifteen to twenty samples that were taken from the microprofiles over an area of about 2 hectares. In each sampling place, the depth of every horizon under study was measured (Table 1). Five to forty composite samples were collected from each stand. The samples were analyzed by different test methods to establish the content of nitrogen, phosphorus, potassium, magnesium, and calcium according to different degrees of solubility. The results were presented both as units of intensity in mg/kg of soil and as units of capacity in kg/ha. The reserve of nutrient elements was calculated on the basis of mean depth of horizons and their average density. The mean values were then calculated for the content of each element in the Ofh, AE, and BvJ horizons in every stand separately, and the average values for all stands of the same type of site.

The average content of nitrogen, phosphorus, potassium, magnesium, and calcium was compared for different soils by different methods within one stand of a particular site, within numerous stands of this site, and within stands of different sites. They were compared by using the variation coefficient (V%) characteristic of the average data. The methods of study and literature discussion connected with solubility problems have been described in a previous study (Ostrowska 1981).

In the soils under study, the content of the nutrient elements, except phosphorus, in the root layer decreased from the Ofh to the BvJ horizon. The content of phosphorus is lowest in the AE horizon. The reserve, however, of all the elements increased from the Ofh to the Bv horizon. The quantity of phosphorus, as well as other elements extracted from the soil, depends on the soil testing method used (Figures 1-7).

Comparison of different soil testing methods for each element shows, on the one hand, highly significant and positive correlation between the individual extractants with similar strengths of extraction, and, on the other hand, no correlation at all between the extractants with different strengths of extraction (Table 2). The Gedrojc soil test was accepted as the most suitable method to determine the easily soluble forms of phosphorus, potassium, magnesium, and calcium in the soils under the pine stands.

The variation coefficient (V%) for average content of a nutrient element in the soil depends on the element and the method of determination. The value of V generally ranged from 20% to 40%, but in some cases was higher (Table 3). The high value of V% in AE horizons reflects the stage of the podzolization process in the soil, while in the Bv horizons it results from variability in the soil properties--especially in the fresh-mixed site, where the mechanical composition of soils is differentiated more than in other sites.

The V% value for average content of phosphorus, potassium, and magnesium, determined by the Gedrojc method, in the soil under an individual pine stand is 30% to 40%; the average for numerous stands is 30% to 60%. Similar values have been found for the reserve of these elements in the soil, excluding phosphorus. The V% values for average

Table 1. Average depth of genetic horizons of soil and their basic characteristics.

Site	Ofh					AE					Bv				
	n	$\bar{x}$	$s_{\bar{x}}$	μ	V%	n	$\bar{x}$	$s_{\bar{x}}$	μ	V%	n	$\bar{x}$	$s_{\bar{x}}$	μ	V%
Under Individual Stands															
Gubin (Bs)	17	2	0	2	0	17	9.8	0.9	8.1-11.5	34	22	20	1.0	18-22	23
Tuchola (Bśw/Bs)	44	3	0.2	2.6-3.4	44	44	8.2	0.3	7.6-8.8	26	44	22	0.7	19-25	22
Janów (Bśw)	36	4	0.2	3-5	25	36	7.8	0.3	7.2-8.3	23	36	20	0.5	18-22	15
Spychowo (Bśw)	41	4	0.2	3.0-5.0	20	41	8.2	0.3	7.4-9.0	22	41	32	0.3	30-34	17
Ruszów (Bśw/BMśw)	22	8.6	0.4	7.8-9.4	19	22	8.0	0.4	7.3-8.7	20	22	33	1.0	31-35	14
Miłomłyn (BMśw)	25	3.1	0.2	2.8-3.4	25	20	1.6	0.5	14-18	10	12	18	0.5	16-20	10
Under Numerous Stands															
Bs	30	1.8	0.8	1.5-2.1	21	30	11	0.7	9-13	33	30	22	1.2	19-25	30
Bśw	240	4.5	0.2	4.0-5.0	49	240	8.0	0.2	7-9	24	240	27	0.6	25-29	29
BMśw	73	3.3	0.2	3.0-3.6	35	73	14	0.6	12-16	31	73	22	0.9	20-24	26

Bs = dry site. Bśw = fresh site. BMśw = fresh-mixed site.

Ofh = raw humus. AE = accumulation-eluvial. Bv = (BvJ Bv) weathering-illuvial and weathering.

n = number of replication. $\bar{x}$ = mean cm. $S_{\bar{x}}$ = mean error cm. μ = confidence interval. V% = variation coefficient.

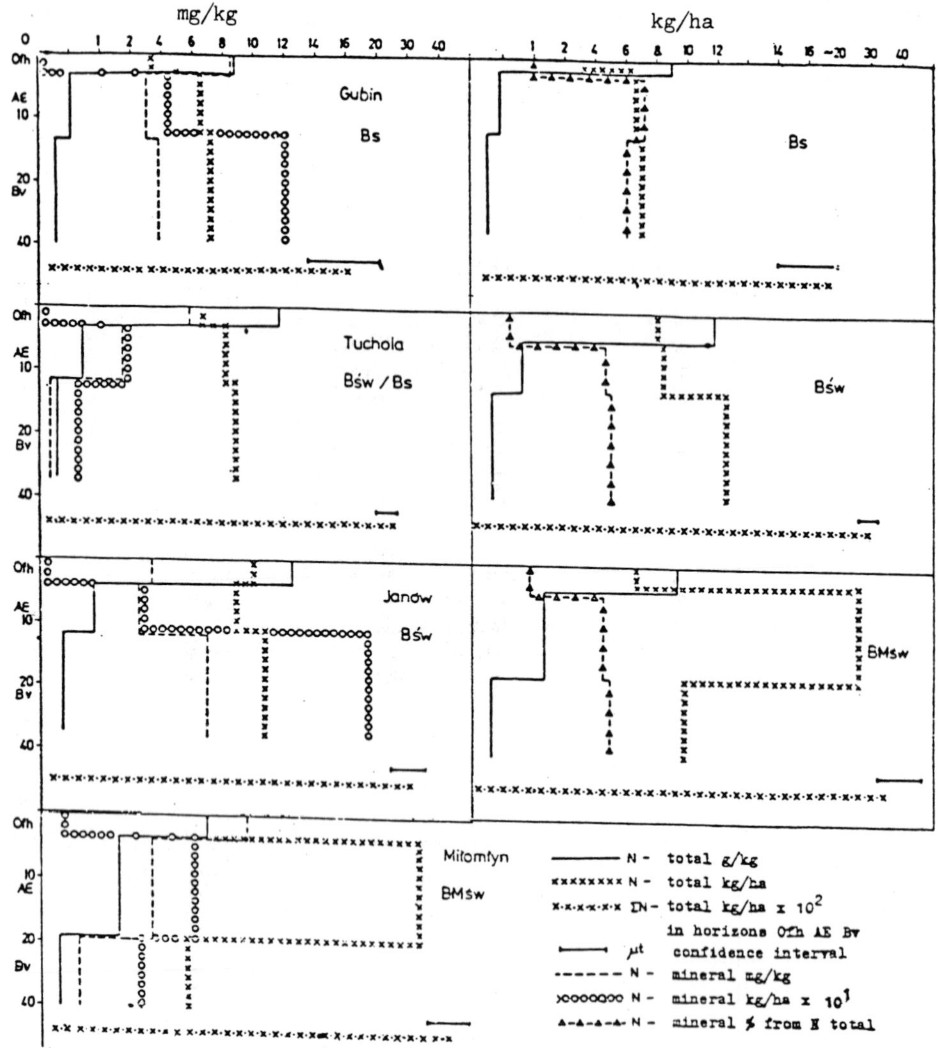

Figure 1. Average content and reserve of nitrogen in soils under individual and numerous pine stands.

Bs = dry site. Bśw = fresh site. BMśw = fresh-mixed site.

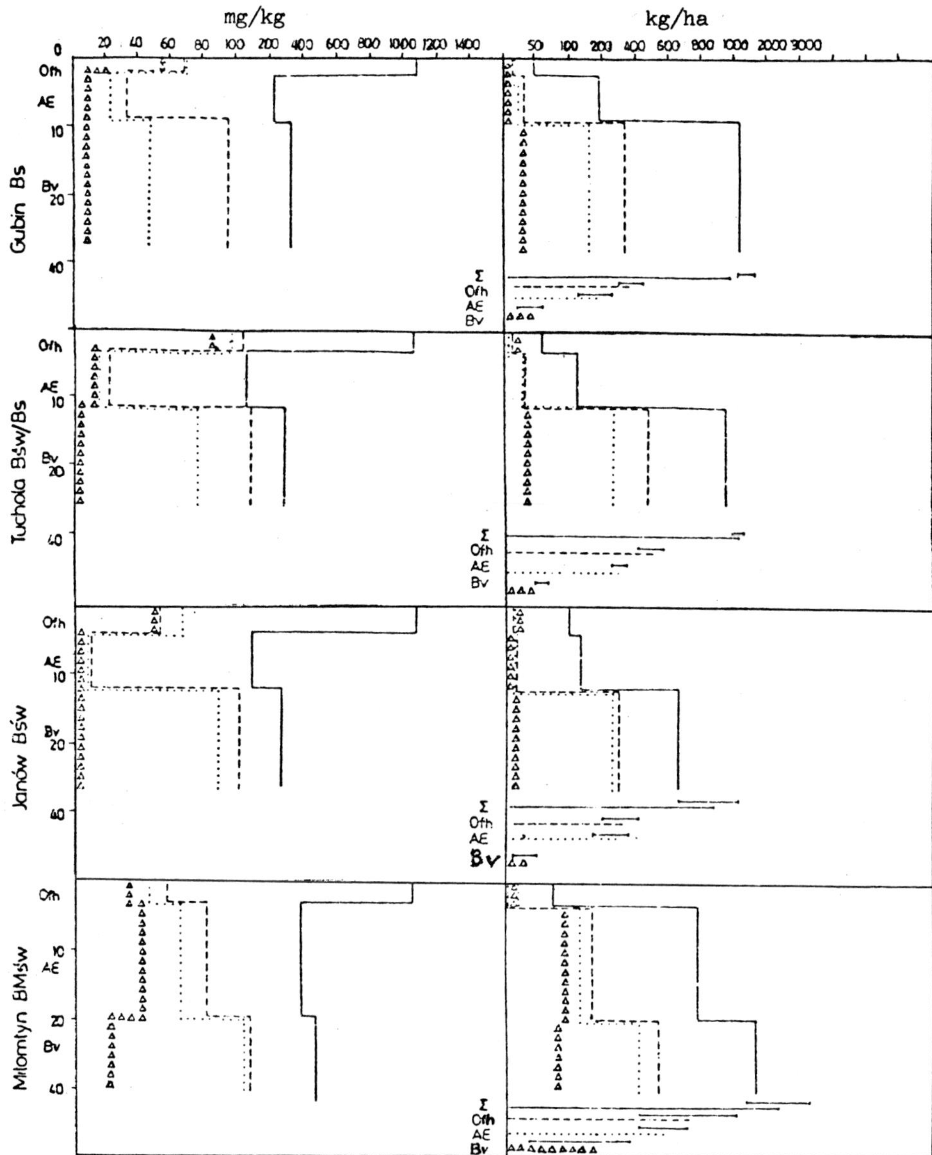

Figure 2. Average content and reserve of phosphorus in soils under individual pine stands.

——— by Bemmelen-Hissink method

-------- by Komprath method

.... by Gedrojc method

ΔΔΔΔΔΔΔΔ by Egner-Riehm method

⊢⊣ confidence interval

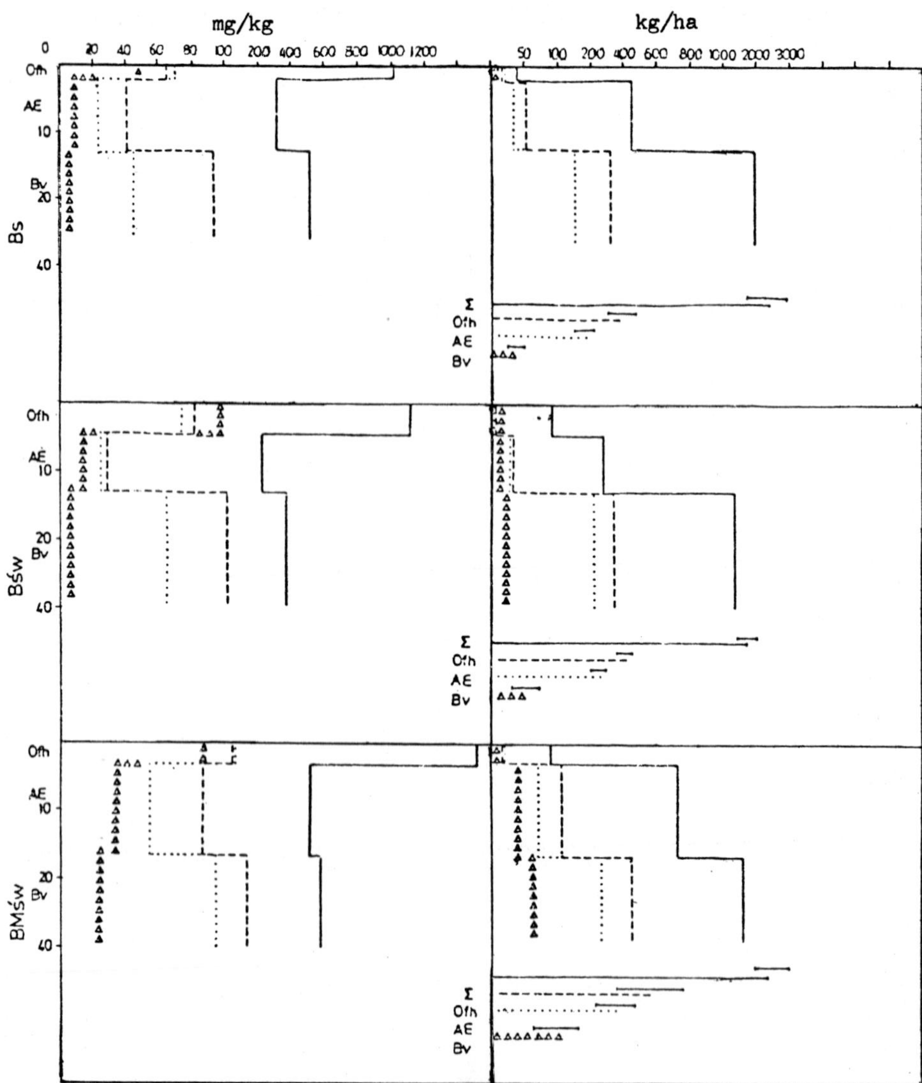

Figure 3. Average content and reserve of phosphorus in soils under numerous pine stands (methods as in Figure 1).

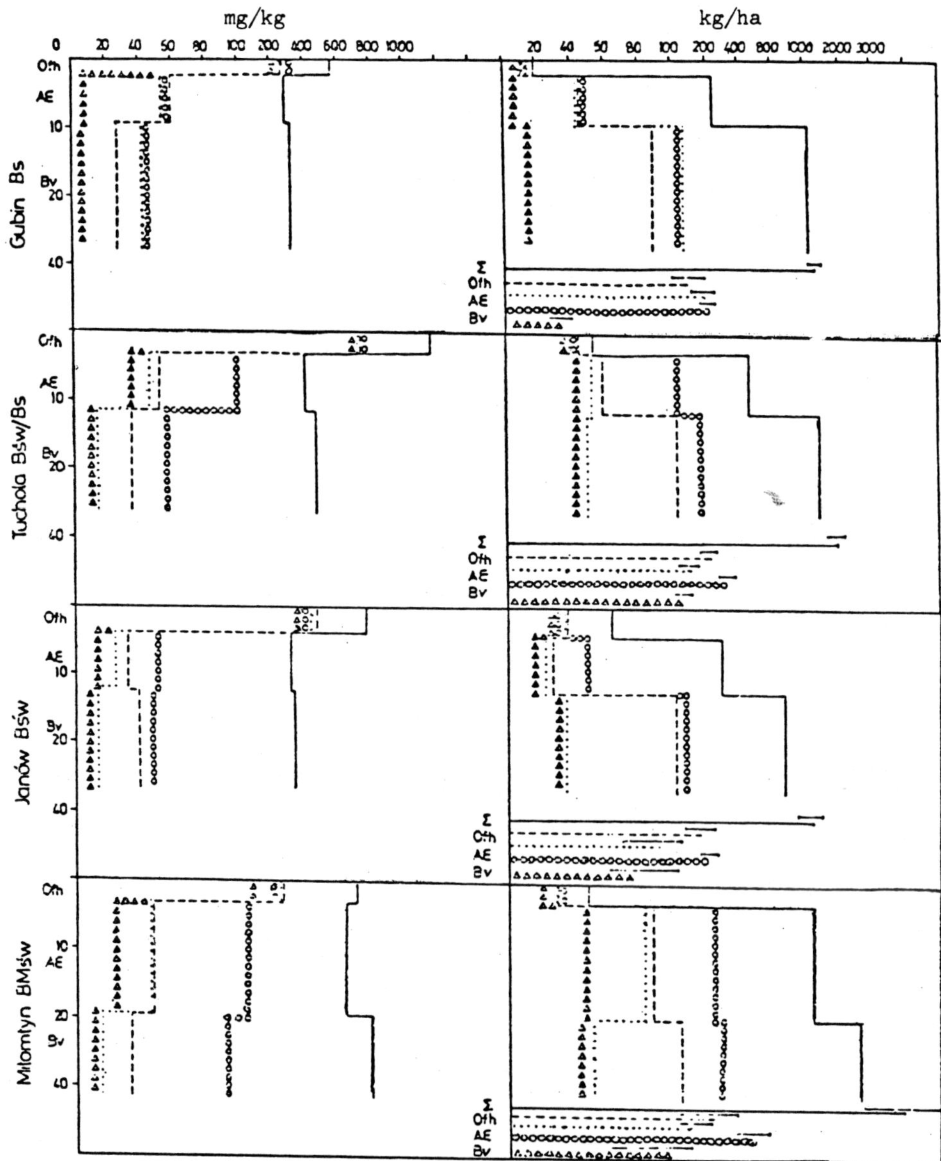

Figure 4. Average content and reserve of potassium in soils under individual pine stands.

——— by Bemmelen-Hissink method
- - - - by Komprath method
. . . . . by Gedrojc method
o o o o o by Greweling-Peech method
ΔΔΔΔΔΔΔΔ by Egner-Riehm method
├─┤ confidence interval

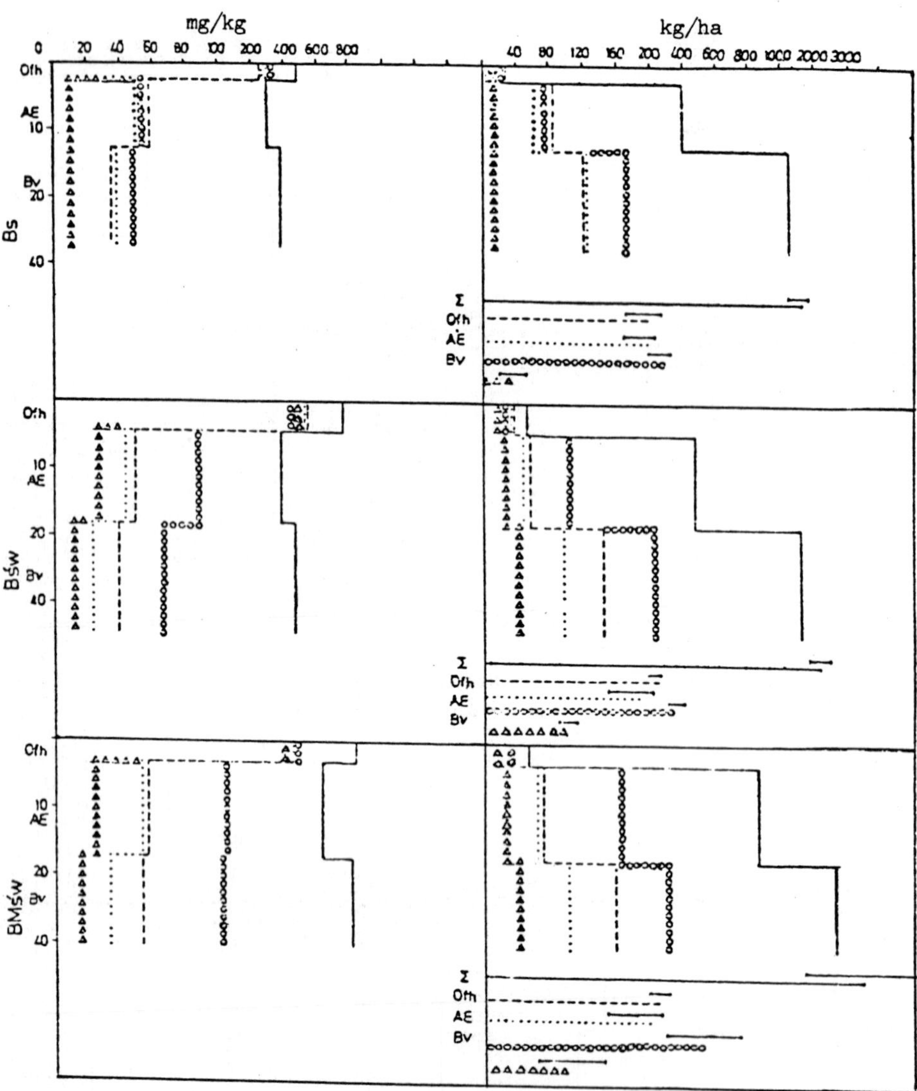

Figure 5. Average content and reserve of potassium in soils under numerous pine stands (methods as in Figure 4).

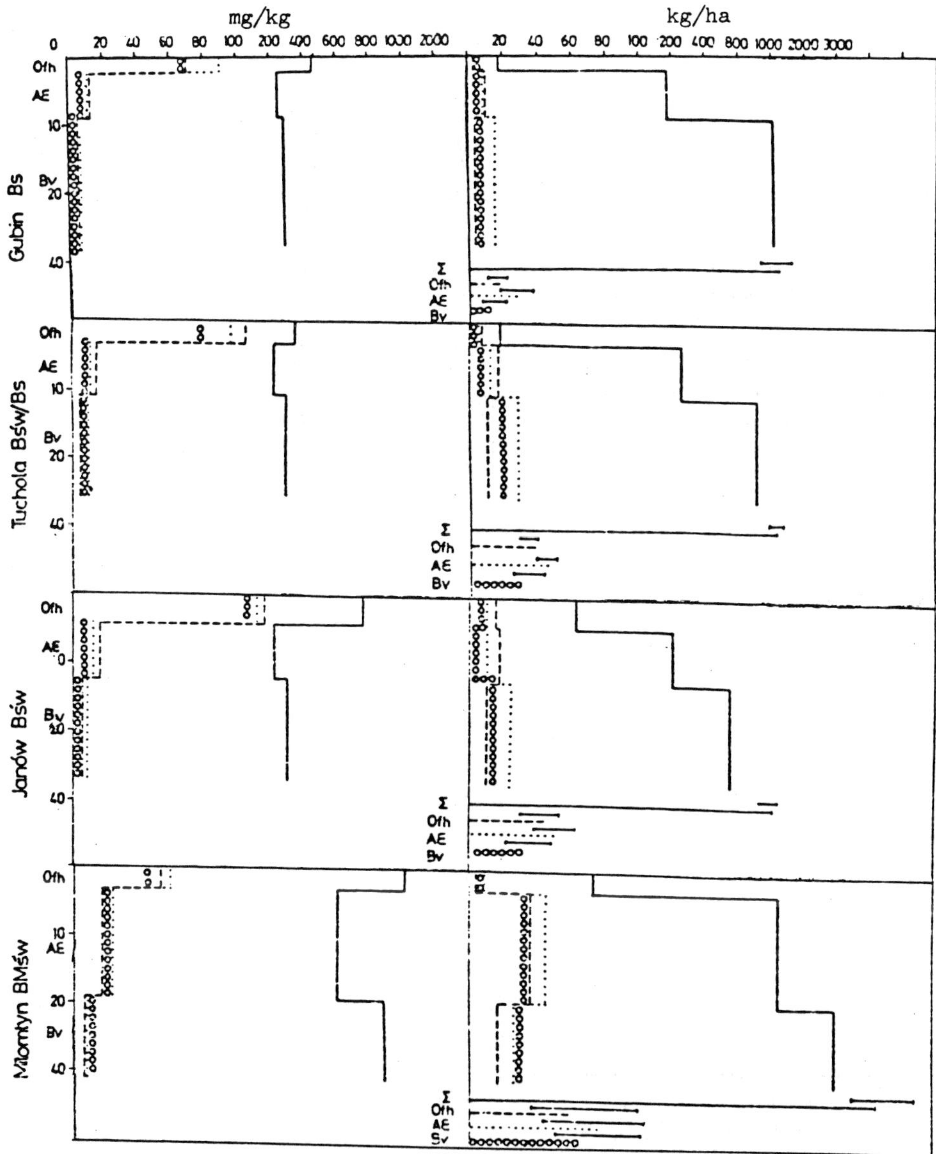

Figure 6. Average content and reserve of magnesium in soils under individual pine stands.

——— by Bemmelen-Hissink method

- - - - by Komprath method

. . . . by Gedrojc method

o o o o by Schachtschabel method

├─┤ confidence interval

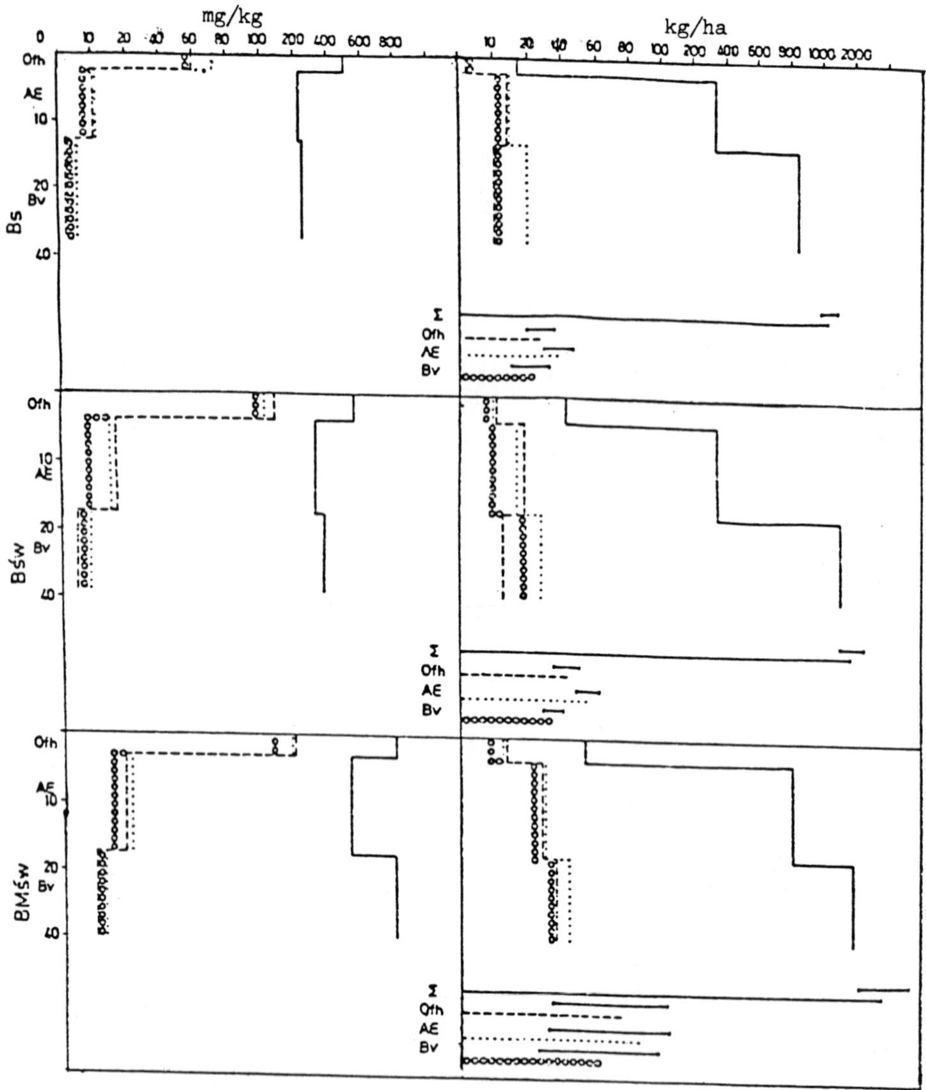

Figure 7. Average content and reserve of magnesium in soils under numerous pine stands (methods as in Figure 6).

Table 2. Correlation of coefficients R^2 for quantities of P, K, Mg, and Ca extracted from soils by various extractants.

Y/X	Horizon	Phosphorus			Potassium			Magnesium			Calcium		
		Ofh	AE	Bv	Ofh	AE	Bv	Ofh	AE	Bv	Ofh	AE	Bv
1.	20% HCl / 5% HCl	0.70**	0.87**	0.48*	0.16	0.90**	0.80**	0.80**	0.85**	0.83**	0.65**	0.48*	0.82**
2.	5% HCl / 1% HCl	0.81**	0.99**	0.93**	0.99**	0.86**	0.75**	0.99**	0.90**	0.34	0.99**	0.94**	0.44*
3.	1% HCl / 0.05N HCl	0.97**	0.99**	0.94**	0.98**	0.54*	0.23	0.95**	0.97**	0.20	0.99**	0.72**	0.79**
4.	20% HCl / 0.05N HCl	0.0	0.0	0.0	0.71**	0.10	0.0	0.0	0.20	0.0	0.47	0.13	0.0
5.	20% HCl / calcium lactate	0.0	0.0	0.0	0.69**	0.0	0.0	--	--	--	--	--	--
6.	20% HCl / 0.1N HNO_3	--	--	--	0.73**	0.26	0.40	--	--	--	--	--	--
7.	20% HCl / 0.025N $CaCl_2$	--	--	--	--	--	--	0.0	0.25	0.54*	--	--	--

1, 2, 3 = R^2 was calculated for 10 soils. 4, 5, 6, 7 = R^2 was calculated for 93 soils.

* Significant for P = 0.05. ** Significant for P = 0.01.

Table 3. Variation coefficient (V%) for average content of N, P, K, Mg, and Ca in soils under pine stands.*

Soil, Site	Genetic Horizon	Nitrogen		Phosphorus				Potassium				Magnesium			Calcium	
		Total	Mineral	B-H	G	K	E-R	B-H	G	G-P	E-R	B-H	G	Sch	B-H	G
						Under Individual Stands										
Gubin	Ofh	31	26	16	49	33	42	19	39	60	46	19	69	55	22	32
(Bs)	AE	43	33	27	41	16	164	14	30	15	40	43	34	16	19	54
	Bv	33	44	19	35	33	66	13	23	10	34	29	58	93	27	80
Tuchola	Ofh	15	49	18	55	53	57	30	37	35	36	13	20	23	36	25
(Bśw/Bs)	AE	26	42	59	35	32	38	16	34	26	42	27	28	30	18	43
	Bv	21	66	31	32	37	--	21	34	15	30	21	25	14	42	50
Spychowo	Ofh	17	22	26	35	42	43	28	30	30	32	29	49	52	38	31
(Bśw)	AE	25	41	58	106	108	105	30	48	44	50	59	41	50	41	33
	BV	35	64	64	34	43	260	31	60	32	32	53	34	30	28	103
Miłomłyn	Ofh	37	53	21	18	19	30	12	10	12	15	21	19	18	10	40
(BMśw)	AE	32	40	14	72	53	56	11	44	14	41	14	63	91	39	55
	Bv	29	147	20	12	16	75	20	16	16	11	15	33	17	22	46
						Under Numerous Stands										
Bs	Ofh	26	26	20	43	25	45	20	37	45	43	26	66	58	30	37
	AE	39	32	66	46	67	97	17	28	26	119	42	36	72	24	75
	Bv	39	51	58	39	50	80	28	37	29	93	36	43	81	26	54
Bśw	Ofh	17	68	24	61	64	69	34	52	52	58	43	47	48	54	38
	AE	34	63	66	100	94	94	34	46	34	61	55	50	90	55	67
	Bv	41	83	52	52	52	191	32	37	28	40	50	33	45	47	64
BMśw	Ofh	41	91	36	93	80	61	45	49	48	61	57	46	42	56	59
	AE	42	83	55	98	96	82	60	39	43	56	54	52	66	57	71
	Bv	38	93	57	93	90	164	118	58	117	90	97	213	190	72	62

Methods: B-H = Bemmelen-Hissink. G = Gedrojc. K = Komprath. E-R = Egner-Riehm.
G-P = Greweling-Peech. Sch = Schachtschabel.

Extractants: B-H = 20% HCl. G = 0.05N HCl. K = 0.05N HCl+0, 0.25N H_2SO_4.
G-P = 1N HNO_3. Sch = 0.025N $CaCl_2$. E-R = calcium lactate.

*Soils, genetic horizons, sites as in Table 1.

content and reserve of phosphorus were higher, especially in the AE horizons (Table 4).

The variation coefficient for the average content of nitrogen was similar to that of the other elements, but generally V% was higher for the mineral than for total nitrogen (Table 3).

DISCUSSION

The basis for assessing the need for fertilizing the pine stands is the knowledge of the nutrient reserve in the soil root layer, while the basis for establishing the fertilization dose is the knowledge of mean nutrient reserve in a definite area. Mean nutrient reserve of soil in a given area has an error the magnitude of which depends on the inhomogeneity of the soil as well as the sampling procedure and the sample analysis.

The studies reported here have demonstrated that the variation coefficients for the mean content of various soluble forms of nutrients of a soil of a definite type and low differentiation of physical and chemical properties are within the limits of 60%. Moreover, the effect of the number of soils under the different pine stands situated in different regions of Poland on the value of the variation coefficient is fairly insignificant. Since the podzolization process occurs mainly in AE horizons and variability in the mechanical composition of soil occurs especially under stands on young soils, the value of the variation coefficient is raised on these soils.

The plant nutrient supply is determined by the nutrient concentration per unit of soil mass, as well as by the nutrient reserve within the reach of the root system. The concentration influences the nutrient uptake per time unit, while the nutrient reserve depends on the possibility of nutrient uptake in the following vegetation periods.

In the case of forest tree species, both values have a great importance for the nutrient supply of stands. For instance, at the Ofh horizon the nutrient concentration is optimal from the standpoint of absorption by the tree roots, but the reserve at that horizon is not sufficient for the stand to produce maximal biomass.

The depth of tree roots determines the pool of nutrients that can be used by the stands. All factors that limit the growth of the tree roots reduce the pool.

Under climatic and soil conditions prevailing in Poland, the pine roots, in the majority of pine stands, grow in the Ofh, AE, and Bv (BvJ + Bv) horizons; but in soils of low water retention, usually in dry-site stands, the main masses of roots are found in the Ofh and AE horizons.

The studies have demonstrated that reserves of easily soluble forms of nutrients in the root layer are different in the individual genetic horizons. The Bv horizon is the main store of phosphorus (90%), while about 50% of the potassium and magnesium is found there. The remaining quantities are found in the Ofh and AE horizons. The reserve of the easily soluble form of lime in the soils under consideration is determined mainly by the Ofh and AE horizons (Table 5). These results exclude the possibility of assessing the soil reserve under the pine stands in the easily soluble forms of nutrients when basing the diagnosis on the analysis of a single horizon only.

The differentiation in content and reserve of nutrients in the individual horizons of the same soil type in a single profile is appreciably higher than that in analogous horizons of the same soil type

Table 4. Average content and reserve of P, K, and Mg in soil and their basic characteristics.

Soils			Genetic Horizons								
			Ofh			AE			Bv		
			P	K	Mg	P	K	Mg	P	K	Mg
Under one	mg/kg	$\bar{x}$	96	707	95	18	46	11	76	15	8
pine stand		$S_{\bar{x}}$	10	49	4	1	3	0.6	5	1	0.3
of fresh	kg/ha	$\bar{x}$	5	39	13	22	51	13	250	81	27
site		$S_{\bar{x}}$	0.5	4	1.8	2	4	1	19	6	1.5
(no. of	mg/kg	V%	55	37	20	35	34	28	32	34	25
replicates,	kg/ha	V%	52	54	46	39	37	34	43	56	31
n = 30)											
Under sixteen	mg/kg	$\bar{x}$	75	513	122	25	42	14	63	24	8
pine stands		$S_{\bar{x}}$	6	49	7	3	2	1	5	2	0.3
of fresh	kg/ha	$\bar{x}$	6	38	11	31	48	17	221	99	28
site		$S_{\bar{x}}$	0.4	2.5	1	6	3	2	13	9	1
(no. of	mg/kg	V%	61	52	47	100	46	50	52	37	33
replicates,	kg/ha	V%	54	54	75	157	56	86	54	89	46
n = 193)											

$\bar{x}$ = mean. $S_{\bar{x}}$ = mean error. V% = variation coefficient.

P, K, and Mg determined by Gedrojc method.

Table 5. Contribution of genetic horizons to the total reserve of elements in the root layer (%).

Genetic Horizon	One Stand				Sixteen Stands			
	N	P	K	Mg	N	P	K	Mg
Ofh	5	2	28	11	5	2	20	20
AE	40	8	37	29	35	12	26	30
Bv	55	90	35	60	60	86	54	50

on large areas. This is corroborated by our earlier statements that soil samples for assessing mean soil fertility under the pine stands are to be taken and analyzed separately from each genetic horizon that enters into the composition of the root layer. At the same time, while defining the variability of the soils of the same class and generated from the same parent rock, the generalization to larger areas can be used of the analytical results.

It is known that the level of soil fertility depends on the analytical method used to determine nutrient elements in soil. The results obtained by the author have shown the highly positive correlation between the individual extractants having similar strengths of element extraction from the soil. In this connection we can use various methods to determine, for example, the content of easily soluble forms of elements in the soils in order to compare their fertility.

The most important problem is to establish the relation between the available quantity of an element, assessed by a definite method, and the amount of this element that can be utilized by a plant. Thus the choice of method should be based on that relation. In addition, the simplicity of analytical procedure and the possibility of measuring several nutrient elements simultaneously are considerations in judging the suitability of the method in question. Following these criteria, the Gedrojc method was chosen to determine the easily soluble phosphorus, potassium, magnesium, and calcium in soils under the pine stands.

In the second part of this study I will try to delineate the utilization by pine (*Pinus sylvestris* L.) of nutrient elements from the soil.

REFERENCES

Burg, I. 1976. International methods for chemical analysis. Intern. Rapport no. 80.

Kowalkowski, A. 1973. Nowsze kierunki w nawożeniu lasów. Sylwan no. 3.

Król, H., and A. Ostrowska. 1980. Właściwości gleb borów sosnowych na terenie Ostańców Gubińskich. Rocz. Glebozn. 31:101-115.

Król, H., and J. Trześniewska. 1978. Air-water relationships in forest soils in Poland. Polish I. Soil Sci. 11:1-10.

Ostrowska, A. 1977. Diagnostic of pine fertilization: Soil sampling and results interpretation. In: Proc. Symp. Soil as a Site Factor for Forests of Temperate and Cool Zones, pp. 166-171. Zwolen, Czechoslovakia.

__________. 1980. Zasobność gleb w składniki pokarmowe i ich wykorzystanie przez sosnę w świetle oceny potrzeb nawożenia drzewostanów sosnowych. Polskie Tow. Gleboznawcze (Warsaw) 5(40):1-144.

Pritchett, W. L. 1979. Properties and management of forest soils. John Wiley and Sons, New York.

Shumakov, V. S., and M. P. Popova. 1977. About diagnostic methods regarding fertilization on forest soils and stands. In: Proc. Symp. Soils as a Site Factor for Forests of Temperate and Cool Zones, pp. 236-241. Zwolen, Czechoslovakia.

STUDIES ON PINE STAND FERTILIZATION.
II. UTILIZATION OF SOIL NUTRIENTS BY PINE

A. Ostrowska

INTRODUCTION

The availability of nutrients for forest plants is affected by many factors related to the features of both soils and plants. Several relevant studies have estimated nutrient utilization by plants on the basis of nutrient accumulation in the biomass produced during a definite period of plant growth and development (Baker et al. 1974; Miller et al. 1979; Weber 1977).

The degree of nutrient utilization from the soil can also be estimated under controlled conditions of plant growth. For this purpose, pot experiments are widely used. In this study nutrient utilization by pine was investigated during the first and second years of growth with the aid of pot experiments, using the soil horizons taken from pine stands growing on various sites.

MATERIALS AND METHODS

Pine seedlings (*Pinus sylvestris* L.) have been used during the first and second vegetation periods as well as the soils taken from the Ofh (organic), AE (accumulation-eluvial), and Bv (illuvial-weathering and weathering jointly) genetic horizons sampled in pine stands of the dry conifer, fresh conifer, and fresh-mixed forest sites. Properties of soils and sampling techniques were described in the first part of the study. The environment of root growth in the pot experiments is spatially limited by the pot size, so in order to provide conditions for proper growth of pines, pots were used containing at least 17 kg of soil. Soil was mixed with quartz sand to intensify nutrient utilization under deficit conditions and to obtain plant response to compensate for this deficit by fertilization. The soil-sand substrates obtained were differentiated with regard to their fertility with the use of fertilization.

Soils (each of the genetic horizons taken separately) were mixed with sand according to the proportions given in Table 1, and placed in the pots. Soil, sand, and substrates both fertilized and nonfertilized were analyzed for contents of potassium, phosphorous, magnesium, and calcium soluble in 0.05N HCl as well as for total nitrogen. Samples for analyzing were taken from the individual pots after about one month of plant growth.

Pines were sown or planted at the end of April 1978, in quantities of 100 seeds or five one-year-old plants per pot. Seedlings were thinned in June to 40 per pot. The plants were fertilized according to the outline given in Table 1.

Table 1. Effect of soil fertility and fertilization on the yields of dry matter and utilization of P, K, Mg, and Ca by pine seedlings.

Soil from Pine Stands	Genetic Horizon	Ratio Soil to Sand (kg/pot)	pH of Sub-strate	P		K		Plant Reaction to PK Fert.	Mg		Ca		Plant Reaction to MgCa Fert.	Yields of Dry Matter (g/pot) Treatment			
				ppm	Utiliza-tion (%)	ppm	Utiliza-tion (%)		ppm	Utiliza-tion (%)	ppm	Utiliza-tion (%)		0	N	NPK	NPKMgCa
I/Sand		0:17	5.1	3	4	12	4	**	4	6	17	8	0	3.4	--	19.7	17.6
I/Bs Gubin	Ofh	0.2:16	3.9	4	22	15	18	**	6	20	22	9	0	6.5	12.0**	24.7**	25.8
	AE	1:16	4.4	3	10	13	10	**	5	32	19	5	0	4.8	6.0*	26.0**	24.0
	Bv	1:16	4.4	4	4	15	11	**	4	22	8	26	0	3.9	8.2**	23.0**	23.1
I/Bs/Bśw Tuchola	Ofh	0.2:16	4.0	4	22	18	16	**	8	18	56	4	*	11.3	12.5	24.7**	28.0*
	AE	1:16	4.3	3	20	13	30	**	5	21	20	9	0	7.2	17.1**	23.7**	24.6
	Bv	1:16	4.4	5	19	13	17	**	4	23	9	22	0	4.8	12.5**	18.9**	18.9
I/Bśw Jenów Lubelski	Ofh	0.2:16	4.1	6	18	15	20	**	6	15	30	3	**	7.8	6.9	15.5**	22.5**
	AE	1:16	4.3	5	18	12	27	**	5	14	14	13	0	6.6	14.7**	25.7*	21.1
	Bv	1:16	4.4	6	12	15	17	**	6	12	9	20	0	4.6	11.7**	18.8**	16.8
I/Bśw Spychowo	Ofh	0.2:16	4.0	5	18	18	10	**	6	17	40	3	0	7.9	9.2	17.7**	17.0
	AE	1:16	4.1	5	24	12	30	*	5	16	20	7	0	6.7	16.7**	18.7*	20.7
	Bv	1:16	4.4	9	20	13	50	0	4	25	8	32	0	12.8	20.8**	21.8	20.3
II/Bśw Spychowo	Ofh	0.5:16	3.8	6	34	19	30	0	12	17	150	2	--	15.2	24.7**	20.8	--
	AE	3:15	3.7	7	18	10	25	0	5	19	22	8	--	6.8	13.7**	12.6	--
	Bv	3:15	4.2	13	10	12	19	*	5	14	12	10	--	3.4	12.1**	14.6	--
II/Bśw Jastrowie	Ofh	0.4:15	4.7	5	22	21	17	**	8	12	48	3	0	11.2	13.7*	25.0**	27.1
	AE	2:16	4.0	3	12	12	17	*	5	9	19	5	**	5.2	7.9*	8.4	16.8**
	Bv	2:16	4.7	13	5	13	17	0	5	16	8	13	*	2.5	11.7**	11.9	13.3*
II/Bśw Twardogóra	Ofh	0.4:15	3.6	7	22	22	17	**	8	11	49	4	--	10.7	13.7*	24.3**	--
	AE	2:16	3.7	6	15	15	6	**	6	4	19	2	**	4.1	5.6	9.5*	16.1**
	Bv	2:16	4.2	12	12	12	11	0	5	12	8	18	0	2.6	9.9**	12.1	8.7

Table 1. Cont'd.

Soil from Pine Stands	Genetic Horizon	Ratio Soil to Sand (kg/pot)	pH of Substrate	P		K		Plant Reaction to PK Fert.	Mg		Ca		Plant Reaction to MgCa Fert.	Yields of Dry Matter (g/pot)			
														Treatment			
				ppm	Utilization (%)	ppm	Utilization (%)		ppm	Utilization (%)	ppm	Utilization (%)		0	N	NPK	NPKMgCa
I/BMśw	Ofh	0.2:16	4.4	8	15	16	18	**	7	14	48	5	0	9.8	10.9	23.0**	23.2
Miłomłyn	AE	1:16	4.5	8	22	18	28	**	6	20	17	10	0	9.0	17.3**	26.3**	25.1
	Bv	1:16	4.5	11	8	19	11	**	6	10	9	20	0	4.8	13.8**	23.9**	20.9
II/BMśw	Ofh	0.5:15	4.4	12	26	30	17	0	17	12	90	2	--	22.1	24.1	20.6	--
Miłomłyn	AE	3:15	4.3	17	18	23	22	0	12	13	23	9	--	15.2	17.9*	19.4	--

Bs = dry site. Bśw = fresh site. BMśw = fresh-mixed site.
I: Experiments were carried out in 1978; fertilization (mg/pot): N = 240, P = 120, K = 156, Mg = 100, Ca = 120.
II: Experiments were carried out in 1979; fertilization (mg/pot): N = 240, P = 60, K = 30, Mg = 100, Ca = 120.
* Significant difference for P = 0.05.
** Significant difference for P = 0.01.
Significant differences were calculated for treatments N>O, NPK>N, NPKCaMg>NPK.
Content of P, K, Ca, and Mg in substrates was determined by Gedrojc method.

Phosphorus, potassium, magnesium, and calcium (water soluble salts) were applied to all pots once only, on the soil surface, two weeks after rising (or taking of seedlings). Nitrogen was given in two doses: the first one with phosphorous, potassium, magnesium, and calcium; the other after thinning.

During the vegetation period, humidity was maintained at 50% of the maximal capillary water capacity of the substrate. The experiment was laid out with three replicates in two consecutive years, 1978 and 1979.

Physicochemical properties of the substrates differed from those of the sand and soils used for preparation. For comparison, an experiment was carried out using soil-water cultures. Soil was added (in the amounts given in Table 2) to 3-liter pots and irrigated with about 2.5 liters of distilled water. Pine seeds were sown in quartz sand and transplanted, after about two weeks of vegetation, to special pots in a quantity of 30 plants per pot. They were cultivated using the water-culture technique for about four months. The experiment was laid out with three replicates, according to the outline given in Table 2.

Plant material collected during and after the vegetation period (in September) was analyzed for dry matter as well as for nitrogen, phosphorus, potassium, magnesium, and calcium content. Measurements of these nutrients both in the plant material and in the soil extracts have been conducted using the widely known methods.

Nutrient utilization was determined as the percentage relationship of the nutrient quantity accumulated in the plant yield to the quantity of nutrients migrating from the substrate to the 0.05N HCl solution.

RESULTS AND DISCUSSION

Mixing soil with the severalfold larger amount of sand brought about the formation of substrates of a definite fertility level that depends on the concentration of nutrients in the components as well as on the nutrient solubility under the physicochemical conditions of the substrate (Figures 1 and 2).

The quantity of phosphorus, potassium, magnesium, and calcium in the substrates, determined by the Gedrojc method, depends on the soil, its genetic horizon, and the ratio of sand admixture. Generally the content of phosphorus, potassium, and magnesium in the substrates was considerably lower than in the soils, though higher than in sand. The calcium content, however, was generally higher in the substrates than in sand. It should be stressed that the sand used to dilute the soils was rich in calcium soluble in 0.05N HCl. Lowering of calcium solubility and, now and then, of the solubility of magnesium and potassium after mixing sand with some soils, and especially with the Bv horizons, may result from the change in pH or from mutual relations between nutrients, and so on.

Fertilizing with doses of nitrogen, phosphorus, potassium, magnesium, and calcium, thus providing the maximal biomass production under conditions of the experiment, equalized the richness of substrates with regard to the easily soluble forms of phosphorus, potassium, magnesium, and calcium. However, in some substrates, especially those with Bv horizons, the quantities of calcium soluble in 0.05N HCl were similar or even smaller than those in the sand (Figure 2).

The comparison of substrates, both fertilized and nonfertilized, with the soils used indicates that the fertility level of soils is higher than that of substrates, even when the latter were fertilized. It should

Table 2. Soil-water culture experiment: yields of dry matter, uptake of N, P, K, Ca, and Mg by pine seedlings, and percentage of utilization of nutrient elements from soil.

Soils	Treatment	Dry Matter (g/pot)	Uptake (mg/pot)					Utilization (%)				
			N	P	K	Ca	Mg	N	P	K	Ca	Mg
Miłomłyn (BMśw)	0.1 kg Ofh	9.9	112.3	13.1	34.8	11.2	7.0	17	90	99	17	58
	1.0 kg AE	7.2	88.8	4.6	25.6	8.6	4.1	5	8	47	10	10
	1.0 kg AE+N	8.8	172.4	6.9	28.2	8.9	4.7	--	12	52	10	12
	1.0 kg Bv+N	6.6	122.3	3.9	19.8	9.1	3.7	--	3	79	21	41
	1.0 kg Bv+NPK^1	10.3	129.9	10.6	48.9	4.3	4.4	--	--	--	11	49
	0.5 kg Bv+NK^2	7.7	151.3	4.9	43.4	7.3	3.5	--	7	--	33	78
Ruszów (Bśw)	0.2 kg Ofh	6.3	95.8	6.5	28.2	3.5	3.7	4	65	52	4	12
	0.2 kg Ofh+NCa	7.0	115.0	5.9	28.6	7.1	3.8	--	59	52	--	13
	2.0 kg AE+N	7.6	139.0	6.4	28.4	6.6	3.8	--	32	17	4	16
	1.0 kg AE+NP	11.3	160.0	5.4	27.1	14.3	4.5	--	--	34	18	38
	2.0 kg Bv+N	5.1	110.0	1.2	18.9	5.3	1.6	--	1	36	22	16
	2.0 kg Bv+NPCa	7.3	107.5	7.7	21.7	26.2	5.2	--	--	42	--	52

Fertilization (mg/pot): N = 200, P = 25, K^1 = 50, K^2 = 100, $CaCO_3$ = 1,000.
Soils, genetic horizons as in Table 1.

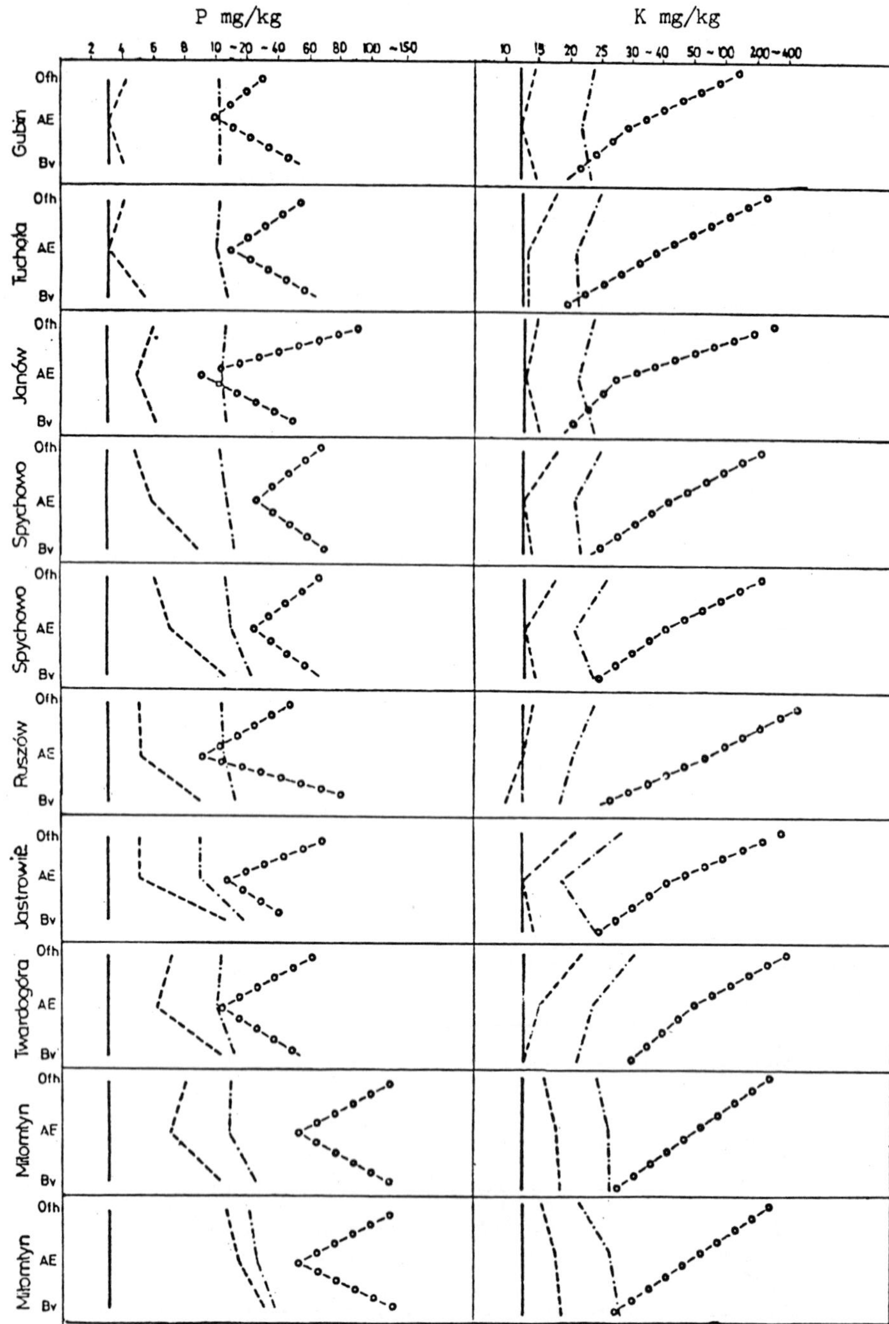

Figure 1. Quantities of P and K extracted by 0.05N HCl from the soils used in experiments. (Fertilization as in Table 1.)

———— sand -------- substrate/sand-soil

o-o-o-o- soil -.-.-.-. substrate/sand-soil/NPKCaMg

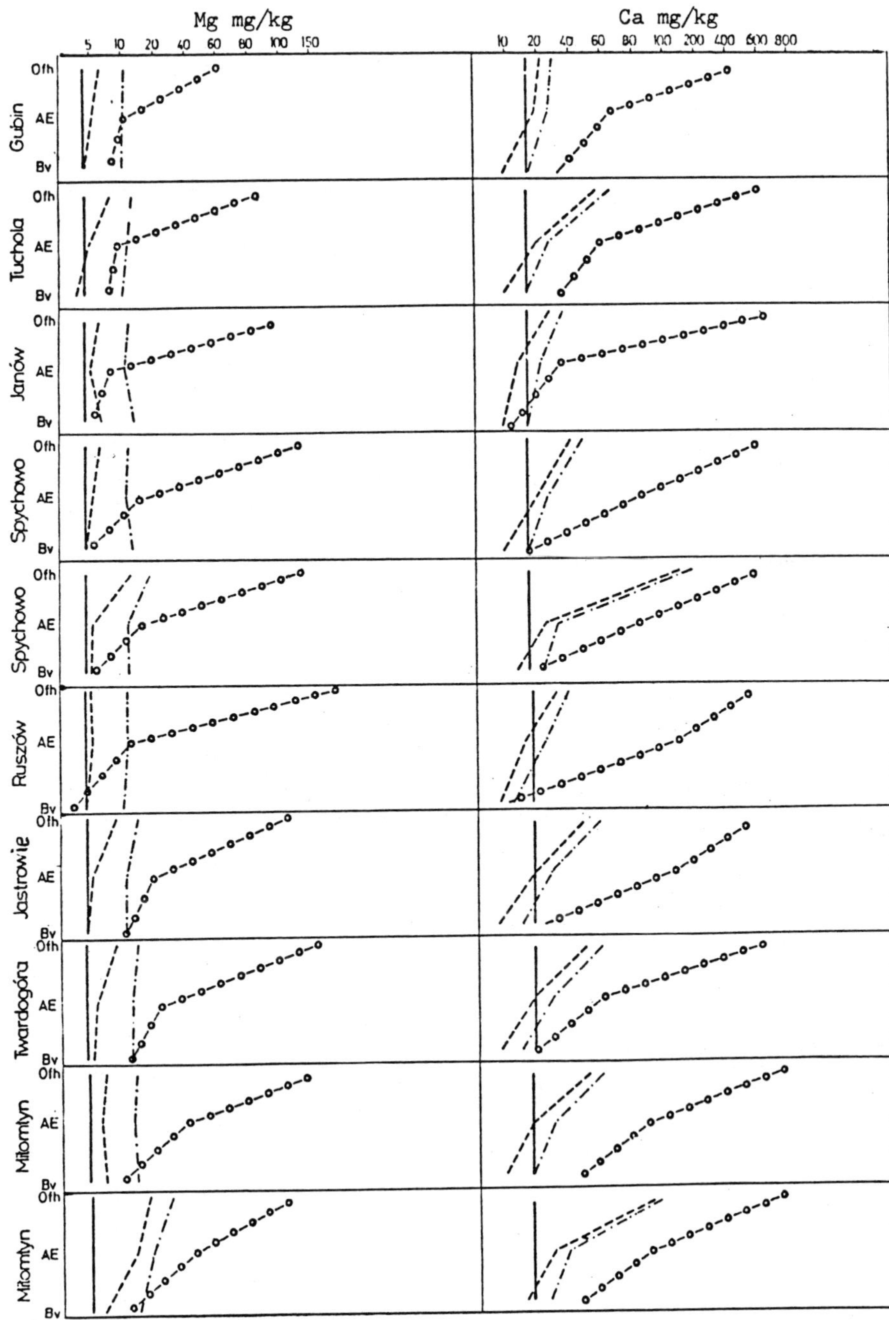

Figure 2. Quantities of Ca and Mg extracted by 0.05N HCl from the soils used in experiments. (Fertilization as in Table 1.)

————	sand	--------	substrate/sand-soil
o-o-o-o-	soil	-.-.-.-.	substrate/sand-soil/NPKCaMg

be stressed, however, that this assessment of soil fertility remains comparative, for it was based solely on the nutritional needs of the pine in pot experiments during the first year of vegetation.

Opinions encountered in the literature that pot experiments are of little assistance when studying forest soil fertility or fertilization needs may be at least partly explained by the fact that no account is taken of the relation between the nutrient demand of plants in a definite period and nutrient pool available in the soil volume of a pot.

Results of the present experiments have shown that sand utilized to dilute soils constitutes a sufficient source of magnesium and calcium for the pines. Plants cultivated on sand did not increase yield of mass after fertilization with the above nutrients (Table 1).

At the soil/sand ratios of 0.2:16 (Ofh) and 1:16 (AE and Bv), the nitrogen fertilization significantly increased yield in the substrates prepared from the AE and Bv horizons in all the soils under study. However, in the substrates with the Ofh horizons the positive effect of nitrogen was visible only in the soil originating from the poorest areas. Joint NPK fertilization essentially increased the mass production in most combinations, when compared with both the nonfertilized combination and the combination fertilized with nitrogen only. Increase of soil amount added to sand brought about, in most cases, the elimination of phosphorus, potassium, magnesium, and calcium fertilization effect (Tables 1 and 2).

The reaction of two-year-old pines grown in the substrates was similar to that of one-year-old pines. The maximal yield of biomass has been obtained after NPK fertilization of the substrates. It is interesting that the amounts of biomass produced by the one- and two-year-old pines are similar in the substrates fertilized with NPK and NPKMgCa. On the other hand, great differences to the advantage of the two-year-old plants were observed in the nonfertilized and nitrogen fertilized combinations. In the above combinations, two-year-old pines utilize more nutrients that one-year-old pines. There occurs here the evident relationship between the plant's ability of nutrient utilization from the soil and biomass production (Table 3). It has been found that under experimental conditions, nitrogen is the element limiting growth in plants. Owing to the above, the degree of phosphorus, potassium, magnesium, and calcium utilization has been calculated, on the basis of uptake of these nutrients from the nitrogen-fertilized substrates. For comparison, the utilization was measured of these nutrients from the substrates prepared with the use of soils sampled in pine stands of the dry conifer (Gubin), fresh conifer (Spychowo), and fresh-mixed (Miłomłyn) forest sites, in all the combinations under study (Tables 3 and 4).

The results obtained have shown that at the level of phosphorus not exceeding 10 ppm in the substrate, utilization of the nutrient amounts to up to 20%, and decreases at higher levels of phosphorus. Utilization of potassium from quartz sand and from most of the soil substrates fluctuated generally between 10% and 20%, whereas the utilization lower than 10% or higher than 20% is related to the biomass yield, and not to the content of potassium in the soil.

Plant responses to fertilization with phosphorus and potassium indicate that at a content exceeding 10 ppm a positive effect of fertilization is rarely observed (Table 1).

The lack of reaction to magnesium fertilization indicates that at a level of 5 to 10 ppm in the substrate, and at utilization reaching up to 10% to 20%, the magnesium level is sufficient to supply the demands of the plants (Tables 1 and 2).

Table 3. Yields of dry matter, uptake of N, P, K, Ca, and Mg by pine, and percentage of utilization of nutrient elements from soil under pine stand of dry site (Gubin).*

		One-year-old Plants												Two-year-old Plants											
		O			N			NPK			NPKCaMg			O			N			NPK			NPKCaMg		
		Ofh	AE	Bv	Ofh	AE	Bv	Ofh	AE	Bv	Ofh	AE	Bv	Ofh	AE	Bv	Ofh	AE	Bv	Ofh	AE	Bv	Ofh	AE	Bv
Dry matter (g/pot)		6.5	4.8	3.9	12.0	6.0	8.2	24.7	26.0	23.0	25.8	24.3	23.1	17.1	11.1	15.7	27.4	26.8	21.1	34.1	28.2	27.6	28.3	29.3	23.0
Uptake (mg/pot)	N	79	57	34	340	162	218	473	446	338	480	438	353	140	82	120	403	255	280	435	321	255	305	298	226
	P	10	6	4	15	5	6	47	41	32	43	35	31	15	11	11	20	15	20	49	39	25	38	34	24
	K	30	19	12	45	23	26	156	146	127	143	150	124	63	46	47	75	56	64	185	161	121	149	135	118
	Ca	13	10	11	14	12	17	38	46	36	36	37	34	27	23	35	62	49	55	45	43	49	32	44	35
	Mg	7	6	5	10	5	8	21	28	16	42	39	36	20	13	18	25	20	24	29	28	29	37	42	34
Utilization from soil (%)	N	4.6	13.4	13.3	--	--	--	--	--	--	--	--	--	8.2	18.6	47	--	--	--	--	--	--	--	--	--
	P	14.7	11.8	5.9	22	9.8	8.8	--	--	--	--	--	--	22.0	21.5	16.2	29.4	29.4	29.4	--	--	--	--	--	--
	K	11.8	8.6	4.7	18	10.4	10.2	--	--	--	--	--	--	24.7	20.8	18.4	29.4	25.3	25.1	--	--	--	--	--	--
	Ca	3.5	3.1	8.0	3.7	3.7	12.5	9.6	15	26	--	--	--	7.2	10.4	25.7	16.5	15.1	40.4	12.0	13.3	36.0	--	--	--
	Mg	6.9	7.0	7.3	7.0	5.9	11.7	20.5	32.9	23.5	--	--	--	19.6	15.2	26.4	24.5	23.5	35.3	28.4	32.9	42.6	--	--	--

* Conditions of experiments as in Table 1.

Table 4. Yields of dry matter, uptake of N, P, K, Ca, and Mg by pine seedlings, and percent of utilization of nutrient elements from soils (Miłomłyn and Spychowo).*

		O			N			NPK			NPKCaMg			O			N			NPK		
		Ofh	AE	Bv	Ofh	AE	Bv	Ofh	AE	Bv	Ofh	AE	Bv	Ofh	AE	Bv	Ofh	AE	Bv	Ofh	AE	Bv
		I Miłomłyn												II Miłomłyn								
Dry Matter (g/pot)		9.8	9.0	4.8	10.9	17.3	13.8	23.0	26.3	23.9	23.2	25.1	20.9	22.1	15.2	4.8	24.1	17.9	7.0	20.6	19.4	7.5
Uptake (mg/pot)	N	129	133	48	264	354	223	392	434	358	400	427	317	516	308	79	608	429	149	536	449	119
	P	16	18	7	21	30	18	50	43	36	50	44	30	33	37	10	54	50	14	56	53	14
	K	37	37	17	49	88	35	171	153	146	170	154	139	85	73	24	92	88	26	97	117	26
	Ca	22	14	12	40	26	36	19	30	31	40	30	31	49	32	8	40	38	16	37	41	16
	Mg	12	11	6	17	20	11	30	22	20	26	39	32	35	24	6	36	27	15	36	29	14
Utilization from soil (%)	N	10	8	20	--	--	--	--	--	--	--	--	--	17	6	12	--	--	--	--	--	--
	P	12	13	4	15	22	9	--	--	--	--	--	--	16	13	2	27	17	3	--	--	--
	K	14	12	5	18	29	11	--	--	--	--	--	--	17	19	6	18	23	7	--	--	--
	Ca	3	5	8	5	9	24	3	10	20	--	--	--	3	8	3	3	10	7	3	10	7
	Mg	10	11	6	14	20	11	26	22	20	--	--	--	13	12	2	12	13	5	12	14	5
		I Spychowo												II Spychowo								
Dry matter (g/pot)		7.9	6.7	12.8	9.2	16.7	20.8	17.7	18.7	21.8	17.0	20.7	20.3	15.2	6.8	3.4	24.7	13.7	21.1	20.8	12.6	14.6
Uptake (mg/pot)	N	119	64	101	198	350	339	370	391	288	312	331	267	379	100	69	704	270	222	571	267	216
	P	12	9	16	15	20	29	46	36	27	33	37	30	27	11	7	35	21	20	47	22	25
	K	43	24	78	33	62	115	135	131	142	111	134	116	74	31	18	96	42	39	92	56	47
	Ca	11	11	36	10	28	33	18	23	43	22	23	31	38	13	10	58	29	19	40	26	37
	Mg	9	8	12	7	14	16	18	15	18	27	33	28	20	9	5	34	16	12	29	16	15
Utilization from soil (%)	N	6	9	33	--	--	--	--	--	--	--	--	--	7	5	8	--	--	--	--	--	--
	P	14	11	11	18	24	19	--	--	--	--	--	--	27	9	3	34	17	10	--	--	--
	K	14	12	35	11	30	52	--	--	--	--	--	--	23	18	9	30	25	19	--	--	--
	Ca	2	3	26	2	8	24	3	7	28	--	--	--	2	3	5	2	8	9	2	7	18
	Mg	9	10	18	7	16	24	18	18	18	--	--	--	10	11	6	17	19	14	14	19	18

* Conditions of experiment as in Table 1.

The calcium contents in the substrates and in the sand are also high, thus the positive effect observed on some substrates after fertilization with calcium and magnesium results from the side influence of these nutrients on properties of the substrate and not from their deficit (Table 1).

In the experiments discussed above one- and two-year-old pines utilized about 20% of phosphorus, potassium, and magnesium content determined in the substrates by the Gedrojc method (Table 3).

Under conditions of soil-water culture, utilization of nutrients is considerably higher and, in the case of phosphorus, amounts to up to 100% (Table 2). Such a high rate is due to the nutrient stress, in relation to the individual nutrients. The utilization depends also on the mobility of nutrients coming from the soil to solution as well as on the higher migration in the solution compared with the substrates.

The degree of nutrient utilization depends on its content in the soil and on the amount of biomass produced in a specific time. Biomass production may be limited by the the deficit of the nutrient under study, deficit of other nutrients, and by additional factors. In this connection, it is difficult to estimate the possibility of nutrient utilization by pine from particular soils. Generally, the utilization of phosphorus and potassium was highest from substrates containing the Ofh horizons. But it was not a rule, because the utilization of nutrients from soil increases as the biomass increases.

On the other hand, with similar biomass yield in different substrates, the degree of utilization decreases as nutrient content in the soil increases. This has already been mentioned above.

At the soil/sand ratio of 0.2:16 (Ofh) or 1:16 (AE and Bv), the yield of biomass is significantly positively correlated with both the content of phosphorus in all the investigated substrates and the content of potassium in the substrates containing the Bv horizon. The correlation coefficients for remaining nutrients turn out to be nonsignificant (Table 5).

At the increased amount of soil in the substrate the significantly positive correlation coefficients were found solely for the relation between the biomass and the content of phosphorus, magnesium, and calcium in the substrates of AE horizon.

Highly significant correlation coefficients were found for the relation between the phosphorus, potassium, and magnesium content in the substrates and the amount of these nutrients accumulated in the mass yield. The lack of significant correlation between the calcium content in substrates and its uptake is due to a high calcium level in the substrates (Table 5).

Significant but mostly lower correlation coefficients for the relation between nutrient content in the soil and accumulation of nutrients in the biomass yield were also obtained elsewhere (Ballard 1978; Kadeba and Byle 1978; McKee 1978).

In our studies, relation between nutrient contents in the substrates and nutrient accumulation in the unit of mass of the aboveground part of plants shows sigificantly positive correlation only when the relation between nutrient content in the substrate and plant yield is highly significantly correlated. The above data reveal that a significant correlation can be obtained only when there is a great deficit or excess of a nutrient in the environment of growth. Hence even the directly proportionate relation is possible, such as between the nitrogen content of needles and the level of nitrogen used in fertilization (Ostrowska et al. 1982). In this connection, the nutrient content of pine needles,

Table 5. Correlation coefficients (R^2) for the relation between soil fertility, yields of dry matter, and accumulation of nutrient elements in pine seedlings.

Y / X	Genetic Horizon	P ppm	P mg/pot	K ppm	K mg/pot	Mg ppm	Mg mg/pot	Ca ppm	Ca mg/pot
Experiment Carried Out in 1978									
Yields of dry matter	Ofh	0.76**	0.74**	0.42	0.43	0.42	0.42	0.06	0.05
	AE	0.69**	0.69**	0.35	0.35	0.28	0.26	0.17	0.17
	Bv	0.63**	0.63**	0.54*	0.54*	0.15	0.15	0.13	0.12
Total accumulation of elements in yields	Ofh	0.89**	0.88**	0.63**	0.63**	0.54*	0.54*	0.10	0.05
	AE	0.81**	0.81**	0.37	0.38	0.87**	0.88**	0.13	0.17
	Bv	0.64**	0.63**	0.58*	0.58*	0.69**	0.69**	0.15	0.12
Content of element in upper parts of plants (mg/g dry matter)	Ofh	0.58*	0.57*	0.56*	0.56*	0.17	0.16	0.05	0.07
	AE	0.61*	0.61*	0.24	0.24	0.22	0.21	0.08	0.08
	Bv	0.44	0.44	0.38	0.33	0.45*	0.45*	0.21	0.21
Experiment Carried Out in 1979									
Yields of dry matter	Ofh	0.46	0.46	0.44	0.31	0.35	0.35	0.32	0.32
	AE	0.85**	0.77**	0.56*	0.63*	0.82**	0.82**	0.78**	0.78**
	Bv	0.23	0.24	0.12	0.12	0.18	0.18	0.39	0.39
Total accumulation of elements in yields	Ofh	0.66**	0.69**	0.58*	0.45	0.60*	0.60*	0.59*	0.59*
	AE	0.92**	0.89**	0.82**	0.79**	0.84**	0.94**	0.79**	0.79**
	Bv	0.25	0.24	0.50*	0.43*	0.0	0.0	0.07	0.07
Content of element in upper parts of of plants (mg/g dry matter)	Ofh	0.27	0.17	0.31	0.32	0.30	0.39	0.16	0.04
	AE	0.79**	0.77**	0.43	0.32	0.36	0.39	0.50*	0.50*
	Bv	0.54*	0.52*	0.37	0.20	0.26	0.26	0.58*	0.58*

R^2 was calculated for treatments O, N, NPK, NPKCaMg.
* Significant difference for P = 0.05.
** Significant difference for P = 0.01.

which is used in practice to diagnose the fertilization needs of pine stands, does not constitute the direct equivalent of the soil richness. This problem is the object of our further investigation.

Application of the results obtained in the course of pot experiments with the use of pine seedlings in order to assess the amount of nutrients or to estimate the effects of pine-stand fertilization causes certain doubts. These doubts result mostly from the direct quantitative comparison of data obtained under the conditions of pot experiments with those obtained under natural conditions of plant growth (Pritchett 1979).

Utilization of nutrients by pine from the soil under natural conditions of stand growth depends on many factors, such as density of roots in a definite volume of soil, water supply, symbiotic organisms, competing activity of other plants in the stand, and the like. Some of these factors will tend to reduce nutrient utilization, whereas the others will increase the utilization compared with that estimated under conditions of pot experiments. Studies on soil richness in the easily soluble forms of phosphorus, potassium, magnesium, and calcium (Ostrowska 1981), as well as on the reaction of pine stands to fertilization with these nutrients (Janiszewski 1981), indicate that utilization of nutrients by pine growing on different forest sites varies within limits similar to those determined under conditions of pot experiments.

The limiting concentrations of 10 ppm of phosphorus and potassium, and of less than 10 ppm of calcium and magnesium, that determine pine growth in the pot experiments will probably be greater under natural conditions. This is also evident from comparison of soil richness with the nutrient content of pine needles (Szczubiałka 1981). The problem, however, requires further investigation.

REFERENCES

Baker, J. B., G. L. Switzer, and L. E. Nelson. 1974. Biomass production and nitrogen recovery after fertilization of young loblolly pines. Soil Sci. Soc. Am. Proc. 38(6):958-961.

Ballard, R. 1978. Use of the Bray soil test in forestry. II. Determination of cation status. N.Z. J. For. Sci. 8:332-343.

Janiszewski, B. 1981. Wpływ nawożenia mineralnego na przyrost masy drewna w drzewostanach sosnowych różnego wieku. Dokumentacja Ustalenie potrzeb nawożenia mineralnego drzewostanów sosnowych. Biblioteka IBL, pp. 76-89.

Kadeba, O., and I. R. Byle. 1978. Evaluation of phosphorus in forest soils: Comparison of phosphorus uptake extraction method and properties. Plant and Soil 49:285-287.

McKee, W. H. 1978. Slash pine (*Pinus elliottii*) seedlings response to potassium and calcium on imperfectly drained coastal plain soil. Plant and Soil 50:15-24.

Miller, H. G., J. M. Cooper, J. D. Miller, and O. J. L. Pauline. 1979. Nutrient cycles in pine and their adaptation to poor soils. Can. J. For. Res. 9:19-26.

Ostrowska, A. 1981. Zasobność gleb w składniki pokarmowe i ich wykorzystanie przez sosnę w świetle oceny potrzeb nawożenia drzewostanów sosnowych. Polskie Tow. Gleboznawcze (Warsaw) 5(40):1-144.

Ostrowska, A., A. Kowalkowski, and P. Szczesny. 1982. Wpływ siarczanu amonu na niektóre właściwości gleb i skład chemiczny igliwia sosny zwyczajnej. Prace IBL 581:72-112.

Pritchett, W. L. 1979. Properties and management of forest soils. John Wiley and Sons, New York.

Szczubiałka, Z. 1981. Zawartość azotu i składników mineralnych w igłach jako podstawa stanu zaopatrzenia sosny zwyczajnej (*Pinus silvestris* L.) w składniki pokarmowe. Rozprawa doktorska, Biblioteka IBL.

Weber, B. D. 1977. Biomass and nutrient distribution in a young *Pseudotsuga menziesii* ecosystem. Can. J. For. Res. 7:326-334.

CONTRIBUTORS

A. E. Akachuku, Department of Forest Resources Management, University of Ibadan, Ibadan, Nigeria

M. Alcubilla, Chair of Soil Science, University of Munich, Munich, West Germany

K. Arimitsu, Forestry and Forest Products Research Institute, Ibaraki, Japan

B. O. Axelsson, Swedish University of Agricultural Sciences, Section of Forest Ecology, Uppsala, Sweden

R. Ballard, Manager, Forest Management Research, Weyerhaeuser Company, Centralia, Washington

D. W. Cole, College of Forest Resources, University of Washington, Seattle, Washington

F. G. Craig, Forests Commission Victoria, Melbourne, Australia

W. J. B. Crane, Division of Forest Research, CSIRO, Canberra, Australia

P. W. Farrell, Forests Commission Victoria, Melbourne, Australia

D. W. Flinn, Forests Commission Victoria, Melbourne, Australia

S. P. Gessel, College of Forest Resources, University of Washington, Seattle, Washington

J. Gračan, Forest Research Institute, Jastrebarsko, Yugoslavia

D. C. Grey, Saasveld Forestry Research Station, George, South Africa

M. Hamamoto, Research Center, Mitsubishi Chemical Industries Ltd., Tokyo, Japan

N. Komlenović, Department of Ecology, Forest Research Institute, Jastrebarsko, Yugoslavia

A. Kowalkowski, Soil and Fertilization Division, Forest Research Institute, Warszawa-Sękocin, Poland

F. J. N. Kronka, Forestry Institute, C.P.R.N., Brazil.

R. Lea, School of Forest Resources, North Carolina State University, Raleigh, North Carolina

H. Löffler, University of Munich, Munich, West Germany

Y. Mashimo, School of Forestry, Tokyo University, Tokyo, Japan

A. Ostrowska, Soil and Fertilization Division, Forest Research Institute, Warszawa-Sękocin, Poland

E. Paavilainen, Metsäntutkimuslaitos, Finnish Forest Research Institute, Helsinki, Finland

J. Päivänen, Finnish Forest Research Institute, Department of Peatland Forestry, Vantaa, Finland

P. Pietiläinen, Metsäntutkimuslaitos, Finnish Forest Research Institute, Helsinki, Finland

R. J. Raison, Division of Forest Research, CSIRO, Canberra, Australia

J. B. Reemtsma, Forest Experiment Station of the Lower Saxony, Göttingen, Germany

K. E. Rehfuess, University of Munich, Munich, West Germany

C. J. Schutz, Natal Forestry Research Centre, Department of Environmental Affairs, Natal, South Africa

R. O. Squire, Forests Commission Victoria, Melbourne, Australia

M. R. Starr, Finnish Forest Research Institute, Department of Soil Science, Vantaa, Finland

K. Takeshita, Kyushu University, Faculty of Agriculture, Kyushu University Forest, Japan

J. L. Timoni, Tupi Experimental Station, Forestry Institute, C.P.R.N., Brazil

A. Van Laar, Faculty of Forestry, University of Stellenbosch, Stellenbosch, South Africa

M. A. M. Victor, Forestry Institute, C.P.R.N., Brazil

L. W. Vincent, Centro de Estudios Forestales de Postgrado, Facultad de Ciencias Forestales, Universidad de Los Andes, Mérida, Venezuela

G. Yamazoe, Forestry Institute, C.P.R.N., Brazil

INDEX